AF572716

DO NOWS

Math Concept Review and Test Practice

For use with pre-algebra and beginning algebra students

Doug Monteath
Don Volle

DALE SEYMOUR PUBLICATIONS

Copyright © 1989 by Pearson Education, Inc., publishing as Dale Seymour Publications®, Inc., an imprint of Pearson Learning Group, 299 Jefferson Road, Parsippany, NJ 07054. All rights reserved. No part of this book may be reproduced or transmitted in any form or by any means, electronic or mechanical, including photocopying, recording, or by any information storage and retrieval system, without permission in writing from the publisher, except for the Blackline Masters, which may be reproduced for classroom use only. For information regarding permission(s), write to Rights and Permissions Department. Published simultaneously in Canada by Pearson Education Canada.

ISBN 0-86651-464-3
Printed in the United States of America

13 14 15 16 08 07 06 05 04

1-800-321-3106
www.pearsonlearning.com

INTRODUCTION

Definition of "Do Nows"

One of the earliest things a student teacher learns is how difficult it is to get 35 exuberant teenagers settled down to learn math. You hold in your hands a part of our answer to this problem: the sheets we call Do Nows.

A Do Now is simply a quick activity that you give to your students at the beginning of your math class. It is intended to take only 5–10 minutes. Many teachers already use this technique under a variety of other names: quizzes, openers, warm-ups, sponges, even "anticipatory sets." We call them Do Nows because, when we introduce the format, we tell students, "Please **Do** these problems **Now**." Students learn that whenever they walk into the classroom, they will find a Do Now awaiting them, and this means they are to sit down immediately and begin work.

Features of These Do Nows

The Do Nows in this book were originally designed for Algebra One classes. However, they do not require any knowledge of algebra since they review facts, skills, strands, and problem-solving techniques introduced in previous mathematics courses. In selecting items, we were guided by the intent and spirit of the California Mathematics Framework and the California Mathematics Curriculum Standards.

One of the important features of these exercises is that they are modeled after different standardized tests our students encounter. Once students are familiar with a particular format, they will find it easier to concentrate on the mathematics involved. For example, you may recognize the "Quantitative Comparisons" format from the SAT (Scholastic Aptitude Test). This two-column format takes some getting used to, so we have included one for every four pages.

The blackline masters in this book can be reproduced as transparencies for the overhead, or they can be photocopied as student handouts. They were originally devised for use with the overhead projector. The idea is simply to project a Do Now on the wall just before class begins; students can write their answers on standard notebook paper. Some teachers are not comfortable using the overhead projector, but we recommend it highly for this purpose. It saves the trouble (and expense) of getting 35 pages a day photocopied. Additionally, the overhead projection helps focus students' attention on the activity and encourages them to use mental math rather than a pencil to determine the answers.

Suggestions for Use

We see Do Nows as a daily activity. They are like stretching exercises before an athletic event; they are like chord exercises before a recital; they help to clear the brain waves of English, social science, and physical education before math class begins. We think it's essential that a Do Now activity be done every day in the first 10 minutes of class. Used casually or half-heartedly, Do Nows are apt to be ineffective.

We do *not* mean to suggest, however, that you use the Do Nows in this book on a daily basis. Variety is essential to good instruction of any type. If you tried to use these sheets consecutively, each day of the year, they would lose their effectiveness; besides, there aren't enough in this set for a whole school year. What works best for us is to use two or three of these pages a week. On the other days, our Do Now time is devoted to other types of exercises. We might present, for example:

- review problems from homework,
- challenging problems from other resources,
- practice problems for tomorrow's test/quiz,
- most frequently missed problems from yesterday's test/quiz.

A blank Do Now sheet is provided on page vi; use it to create alternatives for this 5–10 minute class opening activity.

Special Tips
It's one thing to have an activity set up for students to do as soon as the bell rings; it's another thing to ensure that they do it. Teachers use different techniques for grading or for student accountability. Some pick students randomly to put work on the board for extra credit. Some have students keep a single sheet of paper on which they write their answers for the entire week's Do Nows; this page is then collected on Friday for points. Other teachers include two or three Do Nows on each quiz or test they give, or even give an entire "Do Now quiz" (with maybe 25 questions for 50 points) once during each grading period of a semester. Since the quiz repeats problems that were done in class, it serves to encourage student participation on a daily basis.

One benefit of Do Nows for the teacher is the extra time they give you. While students are working, you can use these 5 or 10 minutes for the details of classroom management. You can take roll, return papers, and—with a little practice and planning—you can even collect homework, record it, and return it in time to go over questions.

Encouraging Mental Math Skills
We recommend that students *not* be allowed to use calculators on these Do Nows. One of the great strengths of these particular problems is that they challenge students to use mental math and estimation to find easier, more efficient methods of solution than the traditional paper-and-pencil algorithms offer. Consider, for example, problem 1 on page 33:

70/137 is approximately equal to

(A) 0.24 (B) 0.39 (C) 0.49 (D) 0.51 (E) 1.96

We do not want our students using a calculator for a problem of this type; nor, for that matter, do we want them using long division. We want them to reason that the value 70/137 is slightly greater than 70/140, or 1/2, so that 0.51 is the only reasonable answer.

As another example, consider problem 1 on page 11, presented in the two-column comparison format:

$$\frac{(248)(6)(7)}{(2)(3)(4)} \qquad \frac{(42)(248)}{(3)(4)(5)}$$

We want our students, in determining which of these values (if either) is greater, to recognize quickly that 6 x 7 = 42, so the two numerators are equal. They should then see that the first denominator (2 x 12) is less than the second (5 x 12); therefore the first quantity is greater than the second quantity. Without putting pencil to paper even once, students should be able to write A as the correct answer.

As you introduce these Do Nows, you might want to demonstrate examples like those just given to make students aware of the possibilities. Then challenge your students not only to find the correct answer, but to do it using the most efficient method possible. As you follow up, you might occasionally investigate the incorrect answers listed on multiple-choice items, discussing the logical reasons for discarding each one.

In all these activities, we hope you will encourage your students to be creative and, most of all, to find pleasure in the activity of learning.

ANSWER KEY

Page 1 A, A, C, D **Page 2** C, A, D, B **Page 3** B, A, A, C, D **Page 4** C, A, E, D **Page 5** A, A, E, C **Page 6** D, C, C, C **Page 7** B, A, A, A, D **Page 8** B, D, C, D **Page 9** C, D, E, B **Page 10** B, E, B, C **Page 11** A, B, B, D, C **Page 12** E, D, C, A **Page 13** B, C, B, D **Page 14** B, B, C, B **Page 15** A, C, C, B, C **Page 16** E, B, B, B **Page 17** E, B, A, C **Page 18** E, D, E, C **Page 19** B, A, C, B, C **Page 20** C, B, B, B **Page 21** C, A, A, C **Page 22** C, C, C, B **Page 23** C, C, D, C, B **Page 24** C, B, B, A **Page 25** B, E, E, D **Page 26** D, C, A, C **Page 27** B, A, C, B, A **Page 28** A, D, A, B **Page 29** A, B, D, A **Page 30** D, C, D, C **Page 31** C, C, D, B, C **Page 32** E, D, C, B **Page 33** D, E, B, A **Page 34** C, B, A, C **Page 35** B, C, C, A, B **Page 36** A, C, B, C **Page 37** C, B, D, E **Page 38** E, B, A, D **Page 39** A, B, A, A, A **Page 40** E, D, D, A **Page 41** E, C, D, C **Page 42** E, B, B, B **Page 43** B, A, C, B, D **Page 44** A, C, D, C **Page 45** D, E, B, D **Page 46** B, B, E, D **Page 47** A, C, D, A, D **Page 48** A, C, B, E **Page 49** D, B, B, D **Page 50** B, D, D, D **Page 51** C, D, B, D, A **Page 52** B, B, D, D **Page 53** C, D, C, A **Page 54** C, C, E, E **Page 55** A, C, C, B, D **Page 56** B, C, B, D **Page 57** C, D, B, C **Page 58** E, E, E, D **Page 59** A, D, C, D, D **Page 60** A, D, A, D **Page 61** D, D, C, C **Page 62** E, C, B, B **Page 63** B, B, D, A, D **Page 64** C, C, C, C **Page 65** C, A, C, A **Page 66** B, C, C, C **Page 67** C, A, D, B, B **Page 68** D, C, D, D **Page 69** C, B, D, C **Page 70** D, A, C, E **Page 71** C, B, B, A **Page 72** D, B, B, D **Page 73** C, B, B, A **Page 74** C, C, C, B **Page 75** A, C, B, B, C **Page 76** A, C, C, D **Page 77** B, C, C, E **Page 78** D, C, E, C **Page 79** B, D, C, A, D **Page 80** C, C, B, E **Page 81** E, B, B, C **Page 82** D, B, B, D **Page 83** A, A, D, A, D **Page 84** A, B, C, C **Page 85** E, C, E, A **Page 86** C, C, D, B **Page 87** A, B, C, C, B **Page 88** B, C, B, B **Page 89** D, B, D, A **Page 90** E, A, D, A

DO NOW

❶

❷

❸

❹

DO NOW

1. Change the fraction 5/8 to a percent.

   (A) 62.5% (B) 5/8% (C) 58% (D) 160% (E) 85%

2. Find the sum: 1/2 + 0.28 + 1/4

   (A) 1.03 (B) 1.3 (C) 10.3 (D) 1.13 (E) 1.06

3. What is the average of the first 11 positive integers?

   (A) 5 (B) 5.5 (C) 6 (D) 6.5 (E) 7

4. What is the perimeter of the figure below?

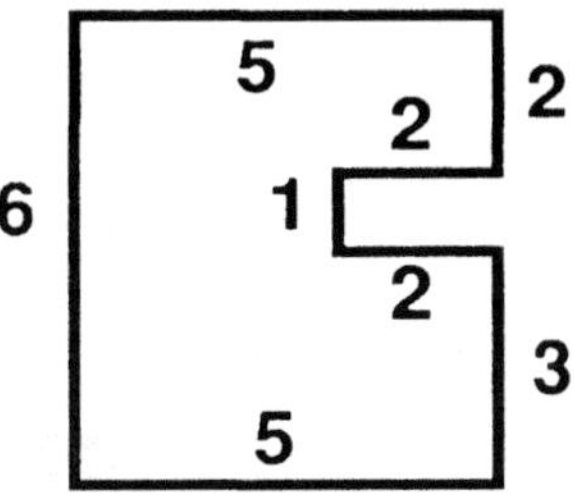

   (A) 34 (B) 30 (C) 28 (D) 26 (E) 24

© Dale Seymour Publications

DO NOW

❶ Divide 24 by 2 2/5.

(A) 57 3/5 (B) 12 2/5 (C) 10 (D) 1/10 (E) 5/288

❷ 25% of 320 =

(A) 80 (B) 128 (C) 240 (D) 1280 (E) 8000

❸ The distance around a circle is its

(A) diameter (B) radius (C) chord (D) circumference (E) area

❹ All sectors of this spinner board are equal in area. What is the probability that, if you spin it once, it will stop on the number 2?

(A) 1/8 (B) 1/4 (C) 1/2 (D) 3/4 (E) 1

© Dale Seymour Publications

QUANTITATIVE COMPARISONS:

Each question below consists of two quantities, one in column A and one in column B. (In some cases, additional information appears centered over the two columns.) You are to compare the two quantities and answer

A if the quantity in Column A is greater;
B if the quantity in Column B is greater;
C if the two quantities are equal;
D if the relationship cannot be determined from the information given.

	COLUMN A	COLUMN B	
❶	(11) (144) (11)	(12) (121) (13)	A B C D ▯ ▯ ▯ ▯
❷	0.03%	$\frac{0.3}{1,000,000}$	A B C D ▯ ▯ ▯ ▯
❸	A single discount of 20%	Two successive discounts of 10% and 10%	A B C D ▯ ▯ ▯ ▯
❹	The area of a square with a perimeter of 32	The area of a rectangle with a length of 16 and a width of 4	A B C D ▯ ▯ ▯ ▯

❺ On a certain test, the average score for the freshmen was 92 and the average score for the sophomores was 88.

	COLUMN A	COLUMN B	
	The average score for the total group	A score of 90	A B C D ▯ ▯ ▯ ▯

© Dale Seymour Publications

DO NOW

1 (3 3/4) (5 1/3) =

(A) 1/20 (B) 45/64 (C) 20 (D) 15 1/4 (E) 64/45

---

2 $(3^2)\ (3^3) =$

(A) 3^5 (B) 3^6 (C) 3^8 (D) 9^5 (E) 9^6

---

3 Which of the following has the greatest value?

(A) 0.05 (B) $(1/2)^2$ (C) 25% (D) 3/10 (E) 0.7/2

---

4 Let $\begin{array}{c|c} x & y \\ \hline w & z \end{array}$ be defined as $\begin{array}{c|c} x & y \\ \hline w & z \end{array} = (x \cdot y) - (y \cdot w)$.

If $\begin{array}{c|c} 5 & 3 \\ \hline a & a \end{array} = 10$, then a =

(A) 2 (B) 3 (C) 4 (D) 5/3 (E) 6

© Dale Seymour Publications

DO NOW

❶ What is 12% of 90?

(A) 10.8 (B) 102 (C) 1080 (D) 7.5 (E) 750

❷ $(0.12)^2 =$

(A) 0.0144 (B) 0.060 (C) 0.144 (D) 0.24 (E) 1.44

❸ Russell's previous test scores are 70, 74, 87, and 85. What score does he need to get on his next test to have an average of 80?

(A) 76 (B) 78 (C) 80 (D) 82 (E) 84

❹ Find the perimeter of the figure if each of the squares has an area of 9 square units.

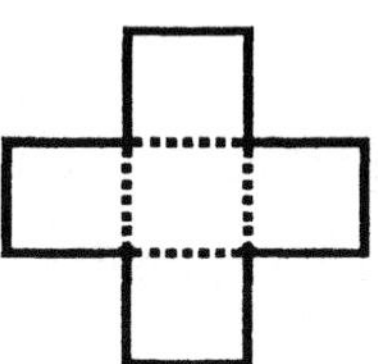

(A) 12 (B) 16 (C) 36 (D) 45 (E) 60

© Dale Seymour Publications

DO NOW

❶ What fraction equals 0.109?

(A) 1/109 (B) 1/10 (C) 10/9

(D) 109/1000 (E) 109/100

---

❷ $(1 - 0.3) - 3.2 =$

(A) 3.9 (B) 2.5 (C) –2.5 (D) –3.9 (E) –10.2

---

❸ What percent of 300 is 24?

(A) 0.08% (B) 1.25% (C) 8% (D) 125% (E) 40%

---

❹ A school employs 12 women, 10 single men, and 8 married men. If 2 single men replace 2 married men, what part of the entire staff is made up of married men?

(A) 5/34 (B) 2/5 (C) 1/5 (D) 5/32 (E) 10/32

© Dale Seymour Publications

QUANTITATIVE COMPARISONS:

Each question below consists of two quantities, one in column A and one in column B. (In some cases, additional information appears centered over the two columns.) You are to compare the two quantities and answer

- A if the quantity in Column A is greater;
- B if the quantity in Column B is greater;
- C if the two quantities are equal;
- D if the relationship cannot be determined from the information given.

	COLUMN A	COLUMN B	
1	(4) (169) (5)	$(13)^2 (5)^2$	A B C D
2	8%	1/125	A B C D
3	The percent increase from $8 to $10	The percent increase from $18 to $20	A B C D
4	The number of edges on a cube	The number of faces on a cube	A B C D
5	$a \cdot b$	$a + b$	A B C D

© Dale Seymour Publications

DO NOW

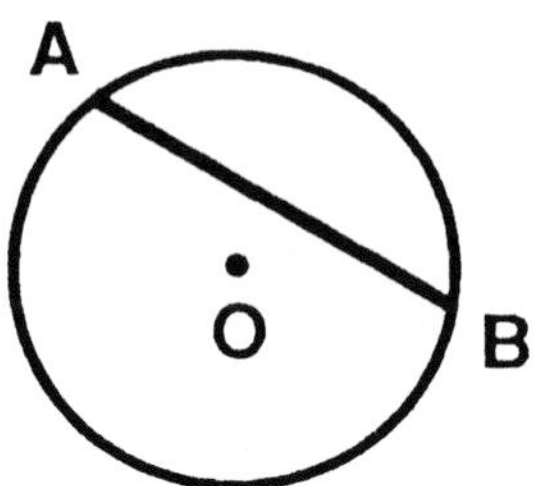

If O is the center of the circle, line AB is a

(A) diameter (B) chord (C) tangent

(D) circumference (E) radius

❷ $\frac{1.52}{0.152} =$

(A) 0.01 (B) 0.1 (C) 1 (D) 10 (E) 100

❸ It takes Mr. Jones 3 hours to clean his house thoroughly. What part of the whole job does he do in 40 minutes?

(A) 1/8 (B) 1/5 (C) 2/9 (D) 1/4 (E) 5/18

❹ The Taft High School basketball team scored 48 points in the first half of a game. In each of the last two quarters, they scored 16 points. What part of their total scoring took place during the second half of the game?

(A) 1/3 (B) 1/5 (C) 1/4 (D) 2/5 (E) 2/3

© Dale Seymour Publications

DO NOW

❶ Forty-eight is two-thirds of what number?

(A) 16 (B) 32 (C) 72 (D) 96 (E) 144

❷ Which of these appears to be an acute angle?

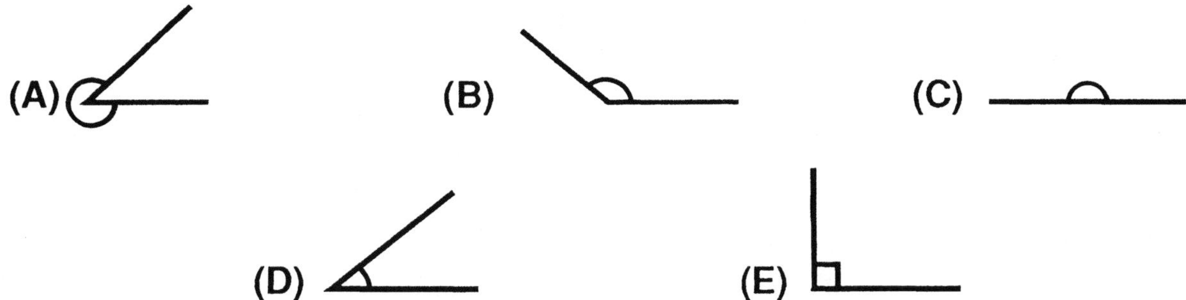

❸ Joe's grocery bill is $7.50, but the clerk deducts $1.00 from his bill for coupons. If Joe gives the grocery clerk $10, how much change should he get?

(A) $8.50 (B) $6.50 (C) $5.50 (D) $4.50 (E) $3.50

❹ At Paso Robles High School, 3/5 of the graduating class is going to college. Of these, 3/5 are going to the community college. What part of the graduating class is going to other colleges?

(A) 3/25 (B) 6/25 (C) 9/25 (D) 16/25 (E) 19/25

© Dale Seymour Publications

DO NOW

❶ 0.12 + 1.25 + 2.3 =

(A) 4.01 (B) 3.67 (C) 3.40 (D) 3.86 (E) 2.87

❷ 21 is what percent of 14?

(A) 1.5% (B) 33 1/3% (C) 60%

(D) 66 2/3% (E) 150%

❸ You pick a card at random from a standard deck of playing cards. What is the probability that it will be a club?

(A) 1/52 (B) 1/4 (C) 1/2 (D) 3/4 (E) 1

❹

A B C D

Distance AC = 8
Distance AD = 10

If distance AB is equal to distance CD, then distance BC is equal to

(A) 2 (B) 4 (C) 6 (D) 8 (E) 10

© Dale Seymour Publications

DO NOW

QUANTITATIVE COMPARISONS:

Each question below consists of two quantities, one in column A and one in column B. (In some cases, additional information appears centered over the two columns.) You are to compare the two quantities and answer

- A if the quantity in Column A is greater;
- B if the quantity in Column B is greater;
- C if the two quantities are equal;
- D if the relationship cannot be determined from the information given.

	COLUMN A	COLUMN B	
❶	$\frac{(248)(6)(7)}{(2)(3)(4)}$	$\frac{(42)(248)}{(3)(4)(5)}$	A B C D ▯ ▯ ▯ ▯
❷	25%	$\frac{1}{0.04}$	A B C D ▯ ▯ ▯ ▯

❸ Erin is 5 times as old as Ryan.
Jennifer is 1/6 as old as Erin.

COLUMN A	COLUMN B	
Jennifer's age	Ryan's age	A B C D ▯ ▯ ▯ ▯

❹ $X \cdot Y \cdot Z = 0$

COLUMN A	COLUMN B	
X	0	A B C D ▯ ▯ ▯ ▯

❺

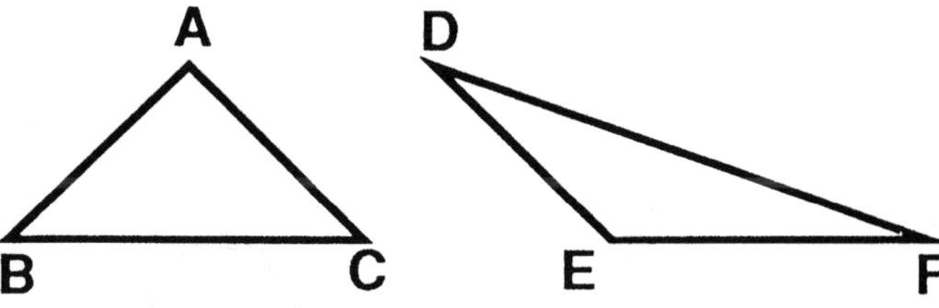

COLUMN A	COLUMN B	
The sum of the interior angles of triangle ABC	The sum of the interior angles of triangle DEF	A B C D ▯ ▯ ▯ ▯

DO NOW

❶ 8 1/3 ÷ 3 1/8 =

(A) 2 5/6 (B) 1 1/3 (C) 2 1/2 (D) 3 5/6 (E) 2 2/3

❷ An item is discounted 20% from its original selling price. If the sale price is $60, what was the original price?

(A) $12 (B) $48 (C) $72 (D) $75 (E) $100

❸ How many fourths of a mile will a bus travel on a 400-mile trip?

(A) 400 (B) 100 (C) 1600 (D) 800 (E) 1/100

❹ In an election, three candidates ran for the same office. One candidate received 3/5 of the votes and another received 2/3 of the remaining votes. What part of all the votes were cast for the candidate who received the least number of votes?

(A) 2/15 (B) 2/5 (C) 1/15 (D) 4/15 (E) 13/15

© Dale Seymour Publications

1. Write 13/25 as a decimal.

(A) 0.13 (B) 0.52 (C) 0.25 (D) 0.38 (E) 1.923

2. 0.6 is what percent of 2.4?

(A) 0.25% (B) 2.5% (C) 25% (D) 250% (E) 25.6%

3. The missing number in 1/8 = ? /0.40 is

(A) 1/200 (B) 1/20 (C) 0.5 (D) 5 (E) 50

4. A 10-gallon can of milk costs $9.50 while a 12-gallon can costs $10.50. How much is saved in the purchase of 120 gallons by buying the larger cans?

(A) $2.50 (B) $3.50 (C) $4.50 (D) $9.00 (E) $7.50

© Dale Seymour Publications

❶ 30 is 60% of what number?

(A) 44 (B) 50 (C) 56 (D) 60 (E) 64

---

❷ An example of a prime number is

(A) 1 (B) 29 (C) 39 (D) 51 (E) 91

---

❸ The Lakers played 60 games of which they won 48. What part of the games played did they lose?

(A) 4/5 (B) 1/4 (C) 1/5 (D) 3/4 (E) 2/3

---

❹ If it costs $160 to carpet a square room 4 yards on each side, how much more would it cost to carpet a square room 5 yards on each side?

(A) $50 (B) $90 (C) $120 (D) $160 (E) $250

© Dale Seymour Publications

DO NOW

QUANTITATIVE COMPARISONS:

Each question below consists of two quantities, one in column A and one in column B. (In some cases, additional information appears centered over the two columns.) You are to compare the two quantities and answer

- A if the quantity in Column A is greater;
- B if the quantity in Column B is greater;
- C if the two quantities are equal;
- D if the relationship cannot be determined from the information given.

	COLUMN A	COLUMN B	
1	$\dfrac{(8-9)(9-10)(10-11)}{(9-10)(10-11)(11-12)}$	0	A B C D
2	0.003	3/1000	A B C D
3	33 1/3% of 8	2/3 of 4	A B C D
4	Sum of any two negative numbers	Sum of any two positive numbers	A B C D

	COLUMN A	COLUMN B	
5	Area of square ACED	Twice the area of triangle DBE	A B C D

© Dale Seymour Publications

DO NOW

❶ Find the sum: 1/3 + 0.7 + 0.25

(A) 21/4 (B) 13/10 (C) 4/15 (D) 0.1275 (E) 77/60

---

❷ If 6 pounds of beans cost $4.80, what is the cost of 10 pounds of beans?

(A) $2.88 (B) $8.00 (C) $14.40 (D) $24.00 (E) $30.00

---

❸ Jack earns $80 for stuffing 4000 envelopes. Sara is hired to stuff 1000 additional envelopes at the same rate of pay. The cost to the company for stuffing all the envelopes is

(A) $80 (B) $100 (C) $120 (D) $200 (E) $1000

---

❹ Let $x^* = x$ if x is zero or positive.
Let $x^* = -x$ if x is negative.
Then $-3^* + 3^* =$

(A) 0 (B) 6 (C) 9 (D) 1 (E) –1

© Dale Seymour Publications

DO NOW

❶ What is 3 1/8 % stated as a fraction?

(A) 1/8 (B) 1/48 (C) 1/24 (D) 2/3 (E) 1/32

❷ The term for the portion of the circle between points A and B is

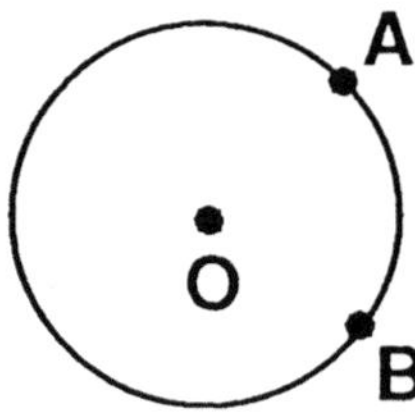

(A) chord (B) arc (C) tangent (D) ray (E) radius

❸ If the ratio of teachers to administrators in a school is 7 to 1, what percent of the staff is teachers?

(A) 87 1/2% (B) 83 1/3% (C) 86%

(D) 83 2/3% (E) 78%

❹ All sectors of this spinner are equal in area. What is the probability that the spinner will stop on an even number?

(A) 1/4 (B) 1/3 (C) 1/2 (D) 2/3 (E) 1

© Dale Seymour Publications

DO NOW

❶ 6/7 of 20 2/3 is

(A) 24 1/9 (B) 23 4/7 (C) 11 1/7 (D) 13 3/7 (E) 17 5/7

❷ Which number is 100 more than 99,999?

(A) 99,099 (B) 100,000 (C) 100,999

(D)100,099 (E) 199,990

❸ Carla needed to compute 40% of 60. All of the following will give a correct answer <u>except</u>

(A) 0.40 (60) (B) 0.60 (40) (C) 0.4 (60)

(D) 60 ÷ 10/4 (E) 60 ÷ 4/10

❹ The toll on a bridge is 75 cents for every car plus 50 cents for every two occupants. At one time a group of cars each containing exactly two occupants passed over the bridge. If the tolls paid totaled $6.25, how many people crossed the bridge in these cars?

(A) 5 (B) 8 (C) 10 (D) 12 (E) 14

© Dale Seymour Publications

DO NOW

QUANTITATIVE COMPARISONS:

Each question below consists of two quantities, one in column A and one in column B. (In some cases, additional information appears centered over the two columns.) You are to compare the two quantities and answer

- A if the quantity in Column A is greater;
- B if the quantity in Column B is greater;
- C if the two quantities are equal;
- D if the relationship cannot be determined from the information given.

	COLUMN A	COLUMN B	
❶	(5) (121) (6)	(112) (52)	A B C D
❷	105% of 300	50% of 600	A B C D

❸ A car travels at an average speed of 30 mph.

COLUMN A	COLUMN B	
The distance traveled in 20 minutes.	10 miles	A B C D

❹ One pencil is drawn from a bag of 3 yellow pencils, 4 green pencils, and 5 blue pencils.

COLUMN A	COLUMN B	
Probability that the pencil drawn is yellow	Probability that the pencil drawn is blue or green	A B C D

❺

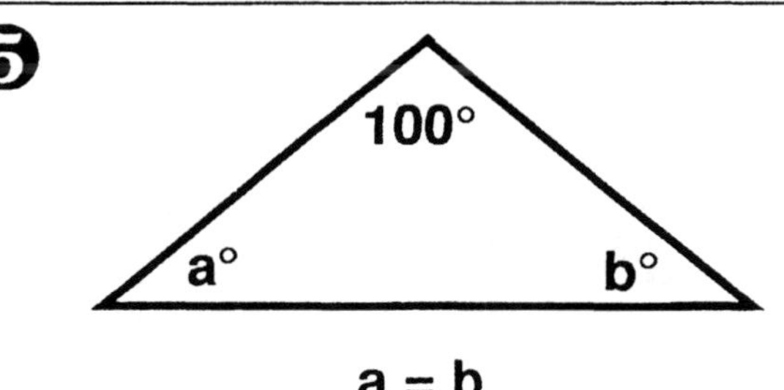

a = b

COLUMN A	COLUMN B	
a°	40°	A B C D

DO NOW

1 (0.4)(0.4)(0.4) =

(A) 0.001 (B) 0.004 (C) 0.064 (D) 0.444 (E) 1.2

---

2 Find the average of 1/2, 3/4, and – 7/8.

(A) 1/24 (B) 1/8 (C) 9/8 (D) 17/24 (E) 5/12

---

3 On a bank account of $3200, how much interest will be paid in one year if the simple interest rate is 5%?

(A) $16 (B) $160 (C) $1600 (D) 64 (E) $640

---

4 Find the average of all the numbers from 1 to 100 that end in 3.

(A) 47 (B) 48 (C) 49 (D) 50 (E) none of these

© Dale Seymour Publications

DO NOW

1 Write 96/256 as a decimal.

(A) 0.96 (B) 0.16 (C) 0.375 (D) 0.556 (E) 0.625

---

2 If T = 16 – 6 • 2 + 4, then the value of T is

(A) 8 (B) 24 (C) 19 (D) 60 (E) 0

---

3 What part of an hour passes between 11:46 PM and 12:10 AM?

(A) 2/5 (B) 7/30 (C) 17/30 (D) 1/6 (E) 1/4

4

Let $\begin{array}{c|c} A & B \\ \hline D & C \end{array}$ be defined as $\begin{array}{c|c} A & B \\ \hline D & C \end{array} = (A - C) + (B - D)$

Then $\begin{array}{c|c} 8 & 6 \\ \hline 2 & 4 \end{array} =$

(A) 4 (B) 6 (C) 8 (D) 10 (E) 12

© Dale Seymour Publications

1. Simplify the following fraction: $\frac{5/9}{1/3}$

(A) 5/27 (B) 3/5 (C) 5/3 (D) 27/5 (E) 3/4

2. Round off to the nearest tenth: 5,316.236

(A) 5,320 (B) 5,316.24 (C) 5,316.2 (D) 5,316.206 (E) 5,316.23

3. How many sixteenths are there in 87 1/2%?

(A) 7/128 (B) 1/14 (C) 14 (D) 128/7 (E) 7

4. The heights of the five starters on Baywood High's basketball team are 5' 10", 6', 6' 1", 6' 4", and 6' 7". The average height of these players is

(A) 6' 1" (B) 6' 2" (C) 6' 3" (D) 6' 4" (E) 6' 5"

© Dale Seymour Publications

DO NOW

QUANTITATIVE COMPARISONS:

Each question below consists of two quantities, one in column A and one in column B. (In some cases, additional information appears centered over the two columns.) You are to compare the two quantities and answer

- A if the quantity in Column A is greater;
- B if the quantity in Column B is greater;
- C if the two quantities are equal;
- D if the relationship cannot be determined from the information given.

	COLUMN A	COLUMN B	
❶	(5) (8) (4) (9)	(72) (10) (2)	A B C D ▯ ▯ ▯ ▯
❷	0.9 times 6	3/5 of 9	A B C D ▯ ▯ ▯ ▯
❸	The total price of a watermelon and a cantaloupe is $2.25.		
	The price of the watermelon	The price of the cantaloupe	A B C D ▯ ▯ ▯ ▯
❹	(square with sides 6 and 6)	(rectangle with sides 5 and 7)	
	Perimeter of the square	Perimeter of the rectangle	A B C D ▯ ▯ ▯ ▯
❺	The height of a flagpole	The length of a football field	A B C D ▯ ▯ ▯ ▯

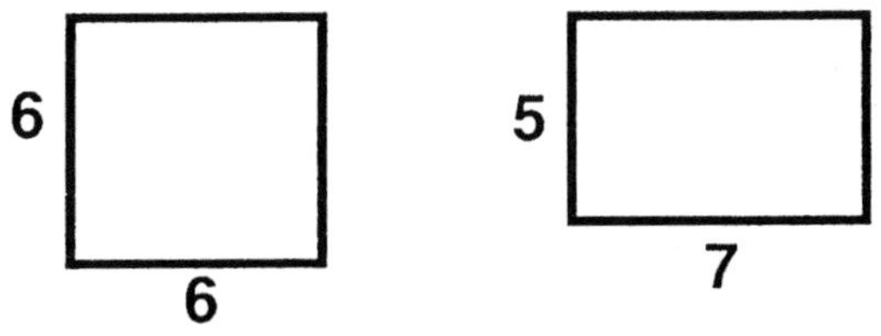

© Dale Seymour Publications

DO NOW

1 Dividing by 0.5 is the same as multiplying by

(A) 1/5 (B) 1/20 (C) 2 (D) 20 (E) 50

2 $\sqrt{53}$ falls between which of the following pairs of numbers?

(A) 6, 7 (B) 7, 8 (C) 8, 9 (D) 26, 27 (E) 52, 55

3 A house is assessed at 20% of its real value. If it is assessed at $30,000, what is its real value?

(A) $60,000 (B) $150,000 (C) $36,000 (D) $100,000 (E) $50,000

4 Find the area (in square units) of the shaded region.

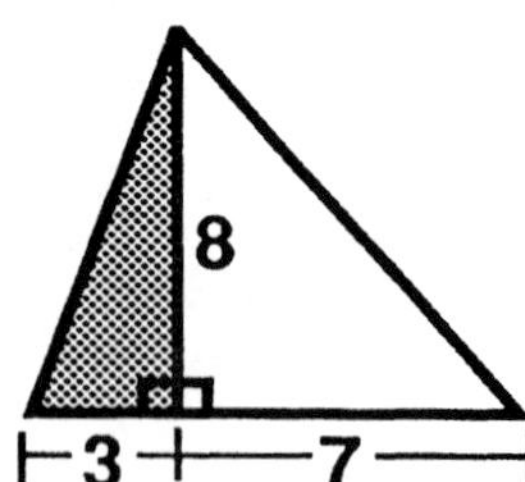

(A) 12 (B) 24 (C) 40 (D) 27 (E) 80

© Dale Seymour Publications

DO NOW

❶ 25% equals how many twenty-fourths?

(A) 4 (B) 6 (C) 8 (D) 24 (E) 50

❷ 5/4 + 1.32 =

(A) 1.57 (B) 2.07 (C) 2.13 (D) 2.53 (E) 2.57

❸ Six men out of a crew of 30 are absent on a particular day. What percent of the crew reported for work?

(A) 15% (B) 20% (C) 25% (D) 75% (E) 80%

❹ Find the area (in square units) of the shaded region.

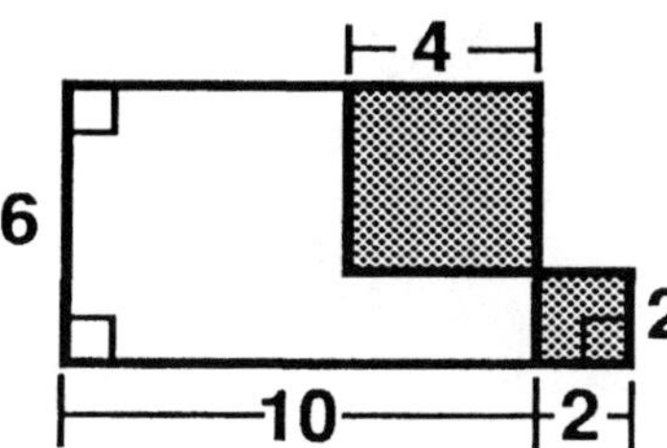

(A) 8 (B) 12 (C) 16 (D) 20 (E) 24

❶ 3/8 + 1/4 − 1/2 =

(A) 1 (B) 3/10 (C) 1/4 (D) 1/8 (E) 0

---

❷ $\frac{3/10}{0.5}$ =

(A) 0.006 (B) 0.06 (C) 0.6 (D) 0.15 (E) 15

---

❸ What is the probability of rolling a two with a standard die?

(A) 1/6 (B) 1/3 (C) 1/2 (D) 2/3 (E) 5/6

---

❹ Mrs. Railsback borrowed $5000 for a year. The cost of the loan was 6 percent of the amount borrowed, to be paid back with the loan at the end of the year. What was the total amount she needed to pay off the loan?

(A) $5003 (B) $5030 (C) $5300 (D) $300 (E) $3000

© Dale Seymour Publications

QUANTITATIVE COMPARISONS:

Each question below consists of two quantities, one in column A and one in column B. (In some cases, additional information appears centered over the two columns.) You are to compare the two quantities and answer

- A if the quantity in Column A is greater;
- B if the quantity in Column B is greater;
- C if the two quantities are equal;
- D if the relationship cannot be determined from the information given.

	COLUMN A	COLUMN B	
1	7/15	29/60	A B C D
2	$\frac{0.5}{0.20}$	0.026	A B C D

3 The Bearcats won 1/2 of their games, lost 1/3 of their games, and tied 1/6 of their games.

	COLUMN A	COLUMN B	
3	The number of games the Bearcats won	The number of games the Bearcats lost and tied	A B C D
4	Degrees in an acute angle	Degrees in an obtuse angle	A B C D
5	A number x plus one	A number x minus one	A B C D

DO NOW

❶ Of the following, which is the closest approximation to the product (0.49) (0.75) (0.665) (0.375)?

(A) 3/32 (B) 3/16 (C) 3/4 (D) 7/16 (E) 5/16

❷ Which of the following is a pentagon?

(A) □ (B) △ (C) ○

(D) ⬠ (E) ⬡

❸ When a coin was tossed, it came up heads 36 times and tails 24 times. What percent of the time did it come up tails?

(A) 40% (B) 66 2/3% (C) 8%

(D) 4% (E) 20%

❹ On a blueprint drawn to the scale of 1 : 60, the length of a building is 14 2/5 inches. The actual length (in feet) of this building is

(A) 7.2 (B) 72 (C) 120 (D) 720 (E) 864

© Dale Seymour Publications

DO NOW

❶ 1 3/4 ÷ 3/11 =

(A) 6 5/12 (B) 7 5/12 (C) 12/77 (D) 21/44 (E) 2 2/21

❷ Which of the following does not equal the others?

(A) 9/25 (B) 3/5 (C) 3 3/25 (D) $\frac{\sqrt{81}}{5^2}$ (E) $\frac{1}{25/9}$

❸ You pick a card from a standard deck of playing cards. What is the probability that it will not be a heart?

(A) 1/52 (B) 1/4 (C) 1/2 (D) 3/4 (E) 1

❹ If a furnace uses 80 gallons of oil in two weeks, how many gallons does it use in 21 days?

(A) 120 (B) 8 (C) 56 (D) 800 (E) 116

© Dale Seymour Publications

DO NOW

❶ If two lines are coplanar and never intersect, these lines are

(A) perpendicular (B) intersecting (C) skew

(D) parallel (E) similar

---

❷ What would be the closest approximation to $\sqrt{66}$?

(A) 6.6 (B) 7.9 (C) 8.1 (D) 8.7 (E) 33

---

❸ A recipe requires 26 oz of sugar and 36 oz of flour. If only 20 oz of sugar are used, how much flour, to the nearest ounce, is needed?

(A) 26 (B) 46 (C) 48 (D) 28 (E) 30

---

❹

◗✢✢◆◆◆☆☆☆☆✻✻✻✻✻◗✢✢◆◆◆

Which of the following will best continue the pattern of the design?

(A) ◆☆ (B) ☆✻ (C) ☆☆ (D) ✻✻ (E) ☆◗

© Dale Seymour Publications

DO NOW

QUANTITATIVE COMPARISONS:

Each question below consists of two quantities, one in column A and one in column B. (In some cases, additional information appears centered over the two columns.) You are to compare the two quantities and answer

- A if the quantity in Column A is greater;
- B if the quantity in Column B is greater;
- C if the two quantities are equal;
- D if the relationship cannot be determined from the information given.

	COLUMN A	COLUMN B	
❶	1/3 – 1/5	2/15	A B C D ▯ ▯ ▯ ▯
❷	$100 - \frac{10}{0.1}$	0	A B C D ▯ ▯ ▯ ▯
❸	x^5	x^4	A B C D ▯ ▯ ▯ ▯

❹ One blank videotape costs \$5.24 while a package of three tapes costs \$14.79.

COLUMN A	COLUMN B	
88 cents	The amount saved by buying a package of three instead of buying three singles	A B C D ▯ ▯ ▯ ▯

❺

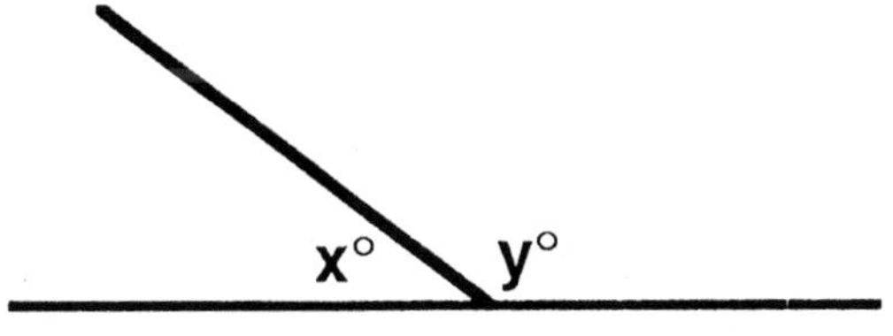

COLUMN A	COLUMN B	
$180° - x°$	$y°$	A B C D ▯ ▯ ▯ ▯

© Dale Seymour Publications

1 The Celtics played 80 games one season and won 48 of them. What percent of the games played did they win?

(A) 24% (B) 32% (C) 40% (D) 48% (E) 60%

---

2 Which of the following is greater than 1/8?

(A) 0.08 (B) $\left(\frac{1}{8}\right)^2$ (C) $\frac{1}{9}$

(D) $\frac{1}{0.08}$ (E) none of these

---

3 40 is 20% of what number?

(A) 8 (B) 32 (C) 200 (D) 400 (E) 800

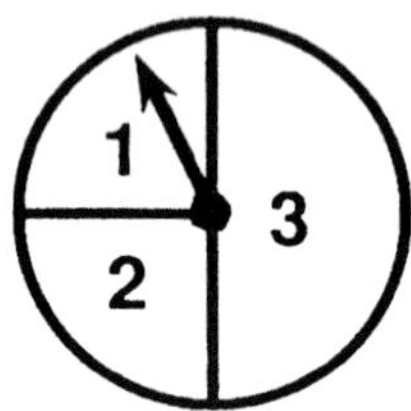

If you spin this spinner, what is the probability that it will stop on 1?

(A) 1/3 (B) 1/4 (C) 1/2 (D) 3/4 (E) 1

© Dale Seymour Publications

❶ 70/137 is approximately equal to

(A) 0.24 (B) 0.39 (C) 0.49 (D) 0.51 (E) 1.96

❷ Which of the following has the greatest value?

(A) 5/12 (B) 4/9 (C) 3/7 (D) 7/13 (E) 3/5

❸ What is the probability that a coin, flipped fairly, will land on heads twice in a row?

(A) 1/2 (B) 1/4 (C) 1/8 (D) 3/4 (E) 0

❹ Shelley spends 1/3 of her paycheck on rent, 3/4 of the remainder on other necessities, and saves the rest. What part of her paycheck does she save?

(A) 1/6 (B) 5/12 (C) 1/2 (D) 7/12 (E) 2/3

© Dale Seymour Publications

DO NOW

❶ Find 30% of 84.

(A) 252 (B) 2.52 (C) 25.2 (D) 2520 (E) 0.252

❷ 320,000 equals

(A) 3.2×10^4 (B) 3.2×10^5 (C) 320×10^2

(D) 3.2×10^{-5} (E) 3.2×10^{-4}

❸ If a car can go 50 miles on four gallons of gasoline, how many gallons will be needed for a trip of 300 miles?

(A) 24 (B) 6 (C) 12 (D) 14 (E) 20

❹ Two-thirds of a class is going on a field trip with the teacher. The students who are not going on this trip are divided equally among 8 other classes. If 3 students are sent to each class, how many students are going on the trip?

(A) 24 (B) 36 (C) 48 (D) 60 (E) 72

© Dale Seymour Publications

DO NOW

QUANTITATIVE COMPARISONS:

Each question below consists of two quantities, one in column A and one in column B. (In some cases, additional information appears centered over the two columns.) You are to compare the two quantities and answer

- A if the quantity in Column A is greater;
- B if the quantity in Column B is greater;
- C if the two quantities are equal;
- D if the relationship cannot be determined from the information given.

	COLUMN A	COLUMN B	
1	$\frac{2/3}{5/6}$	9/10	A B C D
2	(3 + 0.3) (3 – 0.3) (1/4)	(0.25) (3.3) (2.7)	A B C D

3 In a town of 15,000 people, 60% voted. Of those who voted, 60% voted for Mrs. Green.

COLUMN A	COLUMN B	
5400	Number who voted for Mrs. Green	A B C D

	COLUMN A	COLUMN B	
4	x + 1	x – 1	A B C D

5

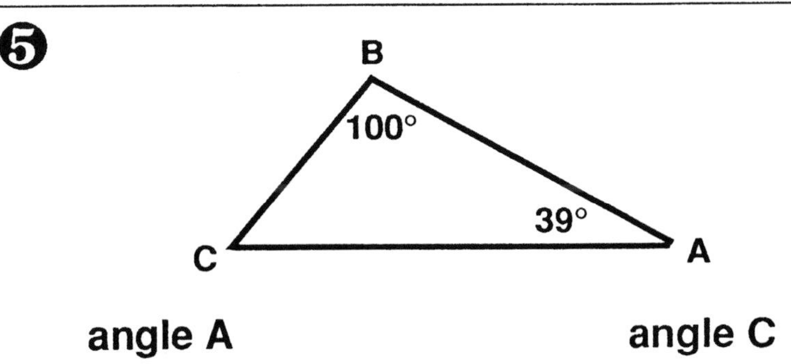

COLUMN A	COLUMN B	
angle A	angle C	A B C D

© Dale Seymour Publications

DO NOW

❶ 4 1/4 + 5 1/3 =

(A) 9 7/12 (B) 7 1/2 (C) 9 1/12 (D) 9 1/6 (E) 20 2/3

❷ What is the average of 1/3 and 1/5?

(A) 1/8 (B) 1/4 (C) 4/15 (D) 8/15 (E) 16/15

❸ Doug had an average of 82 on his first four math tests. After taking the next test, his average dropped to 80. Find his most recent test grade.

(A) 70 (B) 72 (C) 74 (D) 76 (E) 78

❹ If 10 kilometers equal 6.2 miles, then how many miles are in 90 kilometers?

(A) 9 (B) 14.25 (C) 55.8 (D) 59.4 (E) 124

© Dale Seymour Publications

DO NOW

1 Megan buys 8 stamps at 12 cents each, 4 pounds of cheese at \$1.12 per pound and 3 quarts of orange juice at 47 cents per quart. How much change does she get from a 20-dollar bill?

(A) \$6.85 (B) \$8.42 (C) \$13.15 (D) \$14.12 (E) \$14.87

2 Which of the following has the least value?

(A) 1/4 (B) $(0.4)^2$ (C) 1/0.4 (D) $\sqrt{0.49}$ (E) 17/100

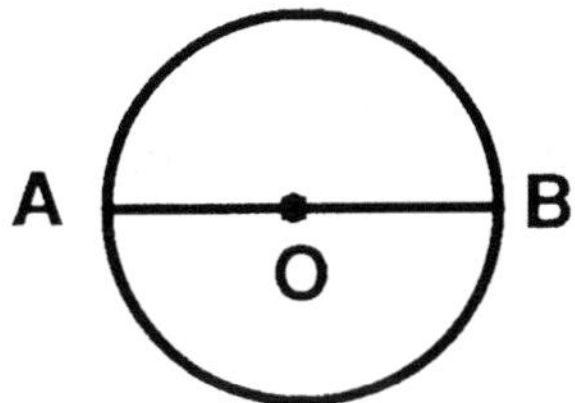

If point O is the center of the circle, which of the following is <u>not</u> true about segment AB? It is

(A) a diameter (B) twice the radius (C) a chord

(D) the circumference (E) equal to AO + OB

4 If John can type 20 pages in four hours, how many hours will it take him to type 70 pages?

(A) 7 (B) 8 (C) 10 (D) 11 (E)14

© Dale Seymour Publications

DO NOW

❶ 0.5% is the same as 1 out of every

(A) 20 (B) 50 (C) 100 (D) 150 (E) 200

❷ A wrestling team consists of 7 seniors, 6 juniors, and 8 sophomores. What percent of this team are <u>not</u> seniors?

(A) 33 1/3% (B) 66 2/3% (C) 50%

(D) 75% (E) 200%

❸ Krista has an average of 82 for three tests. What grade must she get on her next test if she wants to raise her average to 84?

(A) 90 (B) 86 (C) 88 (D) 92 (E) 96

Distance YZ equals distance ZW.
Distance XY equals distance YW.

If distance YZ equals 3, then distance XW equals

(A) 3 (B) 6 (C) 9 (D) 12 (E) 15

© Dale Seymour Publications

DO NOW

QUANTITATIVE COMPARISONS:

Each question below consists of two quantities, one in column A and one in column B. (In some cases, additional information appears centered over the two columns.) You are to compare the two quantities and answer

- A if the quantity in Column A is greater;
- B if the quantity in Column B is greater;
- C if the two quantities are equal;
- D if the relationship cannot be determined from the information given.

	COLUMN A	COLUMN B	
❶	1/4 of 4/5	1/5 of 4/5	A B C D
❷	$(0.2)^2$	$(0.3)^2$	A B C D
❸	3 cubic yards	79 cubic feet	A B C D
❹	The average of 14, 43, and 69	The average of 0, 14, 43, and 69	A B C D

❺

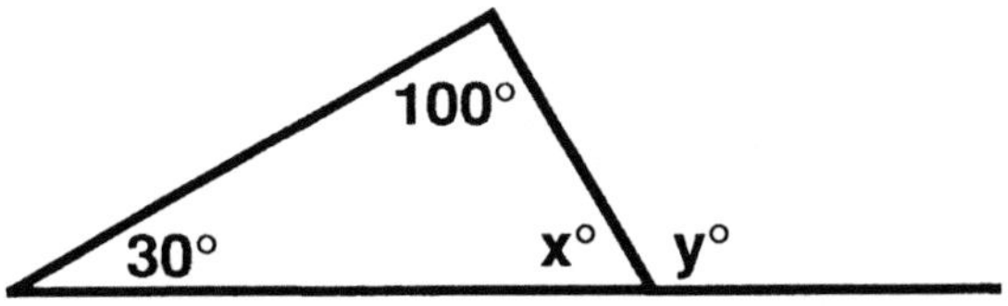

100°
30°
x°
y°

COLUMN A	COLUMN B	
y°	120°	A B C D

DO NOW

❶ (4 7/8) (2 2/3) =

(A) 1/13 (B) 64/117 (C) 8 9/11 (D) 8 7/12 (E) 13

❷ A car averages from 32 to 36 miles per gallon. What is its maximum driving range, in miles, on 12 gallons?

(A) 96 (B) 384 (C) 408 (D) 432 (E) 816

❸ Seth scored an average of 75% on three tests. What score must he get on the fourth test to bring his average to 80%?

(A) 85% (B) 88% (C) 90% (D) 95% (E) 100%

❹

All sectors of this spinner are equal in area. What is the probability that the spinner will stop on a 2 if you spin it once?

(A) 1/3 (B) 1/4 (C) 2/3 (D) 1/2 (E) 3/4

© Dale Seymour Publications

DO NOW

❶

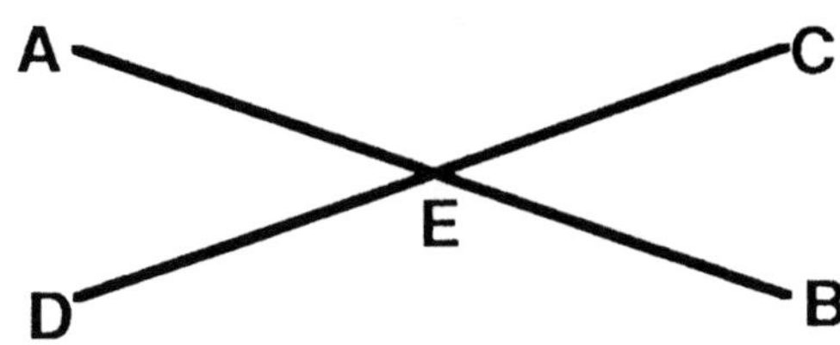

Angle AEC and angle DEB are

(A) adjacent (B) complementary (C) acute

(D) supplementary (E) vertical

❷ An estimate of the value of $\frac{59 \times 2003}{99}$ is

(A) 120 (B) 130 (C) 1200 (D) 1300 (E) 1400

❸ How many buses are needed to carry 196 children to the park if each bus holds 42 children?

(A) 4 (B) 4.28 (C) 4 2/3 (D) 5 (E) 6

❹ For all positive integers n, let

$$n^* = n + 1 \text{ if } n \text{ is even;}$$

$$n^* = n - 1 \text{ if } n \text{ is odd.}$$

If x is a prime number greater than 2, then x^* is

(A) x (B) $x + 1$ (C) $x - 1$ (D) $2x$ (E) $x/2$

© Dale Seymour Publications

DO NOW

1 Which of the following figures is a hexagon?

(A) □ (B) △ (C) ○

(D) ⬠ (E) ⬡

---

2 Uncle Ray purchased 6 pounds of steak priced at $3.19 per pound. How much change did he receive from a 20-dollar bill?

(A) $0.56 (B) $0.86 (C) $2.16 (D) $1.96 (E) $19.14

---

3 Of Templeton High School's graduating class of 60, 24 are boys. Eight boys and 12 girls will <u>not</u> go to college. What percent of the class are girls who will go to college?

(A) 33 1/3% (B) 40% (C) 60%

(D) 66 2/3% (E) 75%

---

4 There are five people in a room. Each shakes hands with everyone else in the room. How many handshakes is this?

(A) 5 (B) 10 (C) 15 (D) 20 (E) 25

© Dale Seymour Publications

QUANTITATIVE COMPARISONS:

Each question below consists of two quantities, one in column A and one in column B. (In some cases, additional information appears centered over the two columns.) You are to compare the two quantities and answer

A if the quantity in Column A is greater;
B if the quantity in Column B is greater;
C if the two quantities are equal;
D if the relationship cannot be determined from the information given.

	COLUMN A	COLUMN B	
1	$\frac{4 + 2/3}{2 - 3/4}$	7	A B C D
2	$\frac{1}{0.05}$	$\frac{1}{5}$	A B C D
3	3 square feet	432 square inches	A B C D
4	The perimeter of a square with an area of 49 square feet	The circumference of a circle with a radius of 5 feet	A B C D
5	$x < 0$ and $y < 0$		
	$x + y$	$x - y$	A B C D

© Dale Seymour Publications

DO NOW

❶ Change 7/200 to a percent.

(A) 3.5% (B) 0.35% (C) 0.035% (D) 28.5% (E) 0.285%

❷ To the nearest tenth of a foot, by how many feet does 12 meters exceed 12 yards? (1 meter = 39.37 inches)

(A) 0.12 (B) 1.2 (C) 3.4 (D) 39.4 (E) 40.4

❸ It is estimated that each person at a picnic will drink 1/4 of a gallon of lemonade. How many gallons of lemonade should be provided if 30 people are expected to attend?

(A) 4 (B) 6 (C) 7 (D) 8 (E) 10

❹ In the series 7, 8, 11, 16, 23, . . . , the next number is

(A) 28 (B) 29 (C) 32 (D) 34 (E) 40

© Dale Seymour Publications

DO NOW

❶ 1.25 + 0.702 =

(A) 0.827 (B) 1.725 (C) 1.927 (D) 1.952 (E) 1.907

---

❷ Which of the following has the greatest value?

(A) 41/49 (B) 23/28 (C) 17/21 (D) 11/14 (E) 6/7

---

❸ A deck of cards contains 52 cards, with 13 cards in each suit. One card is drawn. What is the probability that the card is an ace?

(A) 1/52 (B) 1/13 (C) 4/13 (D) 12/13 (E) 1

---

❹ Marci had exactly $14 before Anna paid her a $52 debt. After the debt was paid, Anna had 1/3 the amount that Marci then had. How much did Anna have before the debt was paid?

(A) $33 (B) $66 (C) $72 (D) $74 (E) $94

© Dale Seymour Publications

DO NOW

❶ Which of these appears to be an obtuse angle?

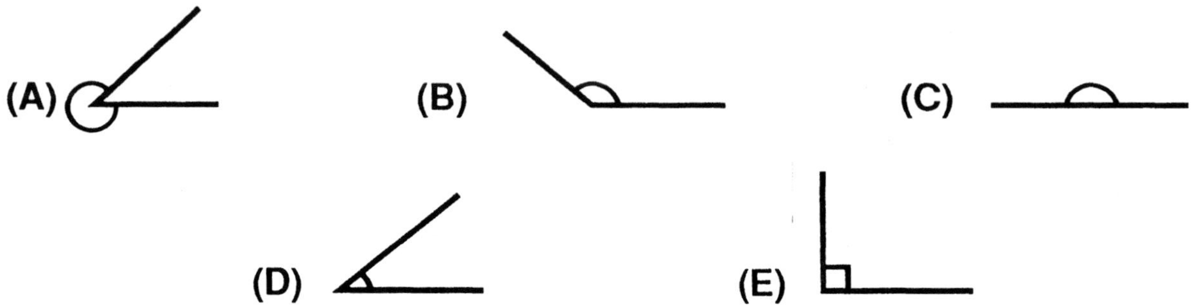

❷ On a test of 25 items, Rachel answered 21 correctly. What percent did she answer correctly?

(A) 0.84% (B) 84% (C) 16% (D) 8.4% (E) 1.6%

❸ A square is divided into four equal smaller squares. If the perimeter of the large square is 2, then the perimeter of one of the small squares is

(A) 1/8 (B) 1/4 (C) 1/3 (D) 1/2 (E) 1

❹ In how many ways can a man seat himself and three other people around a square-topped table if he always wants to sit next to the only mirror in the room?

(A) 2 (B) 3 (C) 4 (D) 6 (E) 12

© Dale Seymour Publications

DO NOW

QUANTITATIVE COMPARISONS:

Each question below consists of two quantities, one in column A and one in column B. (In some cases, additional information appears centered over the two columns.) You are to compare the two quantities and answer

- A if the quantity in Column A is greater;
- B if the quantity in Column B is greater;
- C if the two quantities are equal;
- D if the relationship cannot be determined from the information given.

	COLUMN A	COLUMN B	
❶	$\frac{6}{11} + \frac{1}{17}$	$\frac{8}{17}$	A B C D
❷	$\frac{5}{6}$ of a foot	$\frac{6}{0.6}$ inches	A B C D

❸ The distance from Dave's house to the high school is 3 miles, while the distance from Karen's house to the high school is 4 miles.

	COLUMN A	COLUMN B	
	The distance from Dave's house to Karen's house	5 miles	A B C D
❹	Area of a circle with radius of 2	Area of a square with side of 2	A B C D

❺ $a > b$

	COLUMN A	COLUMN B	
	a^2	b^2	A B C D

© Dale Seymour Publications

DO NOW

❶

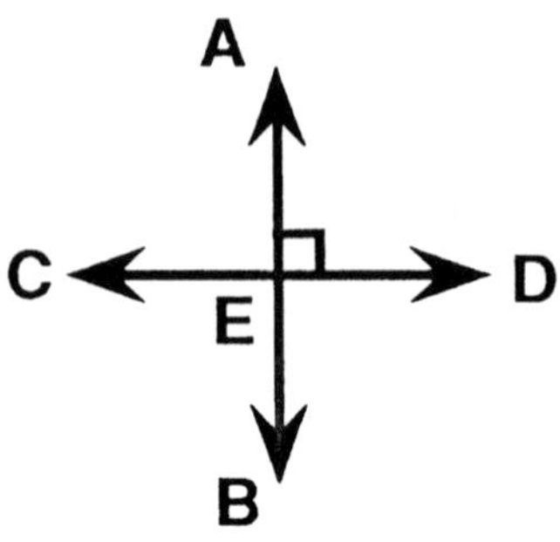

If angle AED is 90°, then lines AB and CD are

(A) perpendicular (B) parallel (C) adjacent

(D) vertical (E) complementary

---

❷ $(1/3)^2 + (4/3)^2 =$

(A) 25/36 (B) 17/18 (C) 17/9 (D) 10/3 (E) 25/9

---

❸ How long will it take a school bus averaging 50 mph to travel 10 miles?

(A) 10 min (B) 12 min (C) 20 min (D) 1 hr (E) 5 hr

---

❹

I

II

III

Which of the above figures can be drawn without lifting the pencil or retracing a segment?

(A) I only (B) II only (C) III only (D) I and II (E) I and III

© Dale Seymour Publications

❶ $\frac{4.04}{0.202} =$

(A) 0.02 (B) 0.2 (C) 2 (D) 20 (E) 200

❷ How many thirds are there in 7/8?

(A) 7/24 (B) 21/8 (C) 8/21 (D) 24/7 (E) 21

❸ There are 36 students in Mrs. Stewart's geometry class. If three-fourths of the students are girls and two-thirds of the boys are less than six feet tall, how many boys in the class are less than six feet tall?

(A) 3 (B) 6 (C) 9 (D) 18 (E) 27

❹

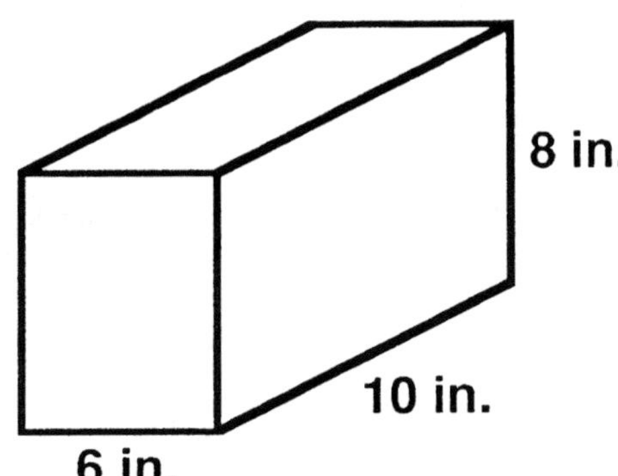

How many cubes 2 inches on an edge can be packed into the box shown above?

(A) 480 (B) 240 (C) 120 (D) 60 (E) 30

© Dale Seymour Publications

DO NOW

❶ 0.0455 =

(A) 91/200 (B) 91/2000 (C) 2/91 (D) 1/455 (E) 1/4550

❷ The fraction 3/8 falls between which of the following pairs of numbers?

(A) 1/5 and 1/4 (B) 0.2 and 0.3 (C) 2 and 3

(D) 0.3 and 0.4 (E) 2/16 and 4/16

❸ Mr. Schleppenbach gives a test with 80 questions and tells his students they must get 48 correct answers to pass. In order to earn a passing score, what is the minimum percent of the questions that must be answered correctly?

(A) 32% (B) 40% (C) 48% (D) 60% (E) 80%

❹ A die is rolled. What is the probability that a multiple of three does <u>not</u> show?

(A) 1/6 (B) 1/3 (C) 1/2 (D) 2/3 (E) 5/6

© Dale Seymour Publications

DO NOW

QUANTITATIVE COMPARISONS:

Each question below consists of two quantities, one in column A and one in column B. (In some cases, additional information appears centered over the two columns.) You are to compare the two quantities and answer

- A if the quantity in Column A is greater;
- B if the quantity in Column B is greater;
- C if the two quantities are equal;
- D if the relationship cannot be determined from the information given.

	COLUMN A	COLUMN B	
1	(0.3) (10,000)	(30) (100)	A B C D
2	x, an even number	y, an odd number	A B C D

3

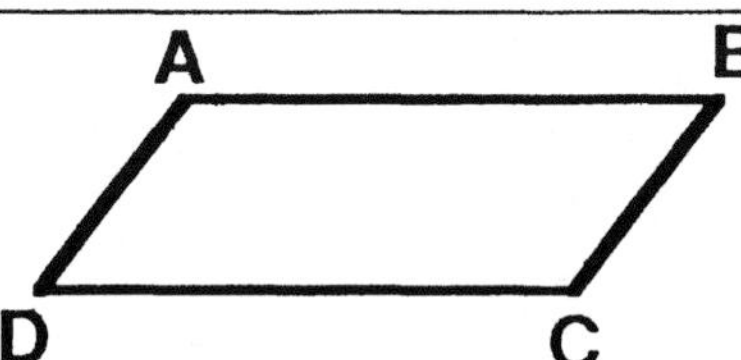

	COLUMN A	COLUMN B	
3	Sum of the angles DAB, ABC, and BCD	360°	A B C D

4

$a < 0$

	COLUMN A	COLUMN B	
4	$a^5 + 1$	0	A B C D

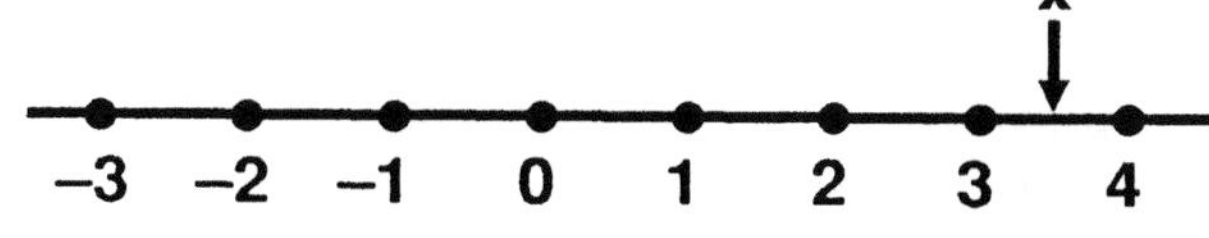

	COLUMN A	COLUMN B	
5	$3 - x$	-1	A B C D

© Dale Seymour Publications

DO NOW

❶ 0.51 + 0.048 + 2.3 =

(A) 1.858 (B) 2.858 (C) 0.788 (D) 7.88 (E) 78.8

---

❷ 2/3 + 1/4 + 1/6 = ? can be most efficiently answered by using a common denominator of

(A) 6 (B) 12 (C) 13 (D) 20 (E) 24

---

❸ You draw one card from a standard deck and you know that it is a face card (jack, queen, or king). What is the probability that it is a queen?

(A) 1/52 (B) 1/12 (C) 1/13 (D) 1/3 (E) 1/4

---

❹ Ms. Patrick has a purse full of coins totaling $7.15. She has at least one of every U.S. coin (dollar, half-dollar, quarter, dime, nickel, and penny). What is the greatest number of quarters she can have in her purse?

(A) 17 (B) 19 (C) 20 (D) 21 (E) 22

© Dale Seymour Publications

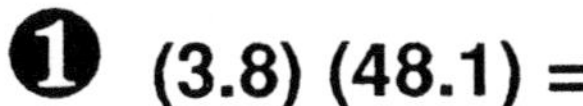

DO NOW

❶ (3.8) (48.1) =

(A) 203.28 (B) 120.18 (C) 182.78 (D) 2005.32 (E) 12.618

❷ 4 + 3/10 + 7/1000 =

(A) 4.037 (B) 4.703 (C) 4.37 (D) 4.307 (E) 4.37

❸ A turkey is to be cooked 20 minutes for each pound. If the turkey weighs 16 pounds and is to be ready for dinner at 6:00 PM, at what time should it be put into the oven?

(A) 11:40 AM (B) 12:20 PM (C)12:40 PM (D) 1:20 PM (E) 1:40 PM

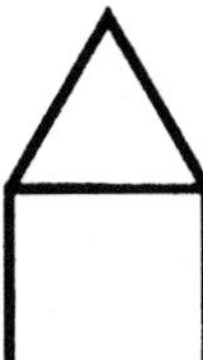

In the above figure, the triangle is equilateral and the area of the square is 64. The perimeter of the triangle is

(A) 24 (B) 32 (C) 40 (D) 48 (E) 64

© Dale Seymour Publications

DO NOW

 What decimal fraction is equivalent to 17/25?

(A) 68.0 (B) 6.8 (C) 0.68 (D) 0.068 (E) 0.17

❷ How much greater is 23 1/8 than 14 3/4?

(A) 8 1/4 (B) 8 5/8 (C) 8 3/8 (D) 9 3/8 (E) 9 5/8

❸

The area of the above circle in square units is

(A) 25 (B) 5π (C) 10π (D) 50 (E) 25π

❹ A box contains 12 jars, some opened and some closed. Each of the following could be the ratio of the opened to closed <u>except</u>

(A) 1/1 (B) 1/2 (C) 2/1 (D) 1/11 (E) 3/4

© Dale Seymour Publications

DO NOW

QUANTITATIVE COMPARISONS:

Each question below consists of two quantities, one in column A and one in column B. (In some cases, additional information appears centered over the two columns.) You are to compare the two quantities and answer

- A if the quantity in Column A is greater;
- B if the quantity in Column B is greater;
- C if the two quantities are equal;
- D if the relationship cannot be determined from the information given.

	COLUMN A	COLUMN B	
❶	4370 rounded to the nearest hundred	4349 rounded to the nearest hundred	A B C D ▯ ▯ ▯ ▯
❷	Claudia's rate if she drives 100 miles in 2 1/2 hours	Sally's rate if she drives 20 miles in 30 minutes	A B C D ▯ ▯ ▯ ▯

❸

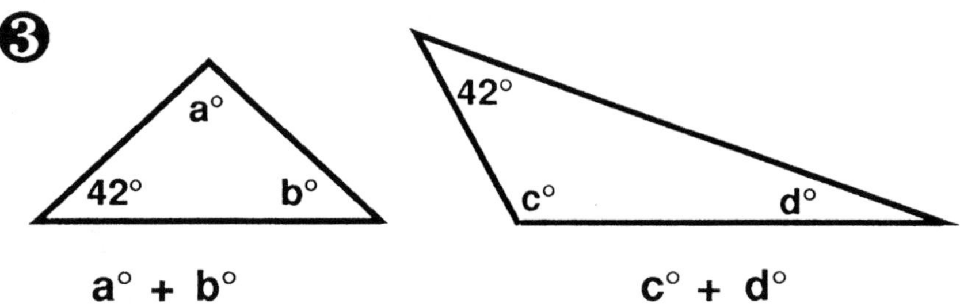

	COLUMN A	COLUMN B	
	$a° + b°$	$c° + d°$	A B C D ▯ ▯ ▯ ▯

❹ The average of six numbers is 25.

	COLUMN A	COLUMN B	
	The sum of the six numbers	160	A B C D ▯ ▯ ▯ ▯

❺ $a < 0$

	COLUMN A	COLUMN B	
	$a + a + a$	$(a)(a)(a)$	A B C D ▯ ▯ ▯ ▯

© Dale Seymour Publications

DO NOW

1. What common fraction is the same as 0.207?

(A) 207/100 (B) 207/1000 (C) 1/207

(D) 27/100 (E) 27/1000

2. What part of an hour passes between 2:44 PM and 3:32 PM?

(A) 0.48 (B) 0.88 (C) 4/5 (D) 3/5 (E) 5/3

3. A box contains 15 red cubes, 12 yellow cubes, and 23 blue cubes. One cube is drawn from the box. What is the probability that the cube is not red?

(A) 4/5 (B) 7/10 (C) 3/7 (D) 7/3 (E) 12/23

4. If the sum of three consecutive odd integers is 417, what is the greatest integer?

(A) 135 (B) 137 (C) 139 (D) 141 (E) 143

© Dale Seymour Publications

DO NOW

❶ How much change would you receive from a 20-dollar bill if you bought items costing \$1.55, \$4.29, and \$5.20?

(A) \$7.76 (B) \$8.22 (C) \$8.96 (D) \$9.36 (E) \$11.15

❷ How many awards can be made from 2 3/4 yards of ribbon if one award can be made from 1/8 yard?

(A) 4 (B) 8 (C) 11 (D) 22 (E) 28

❸

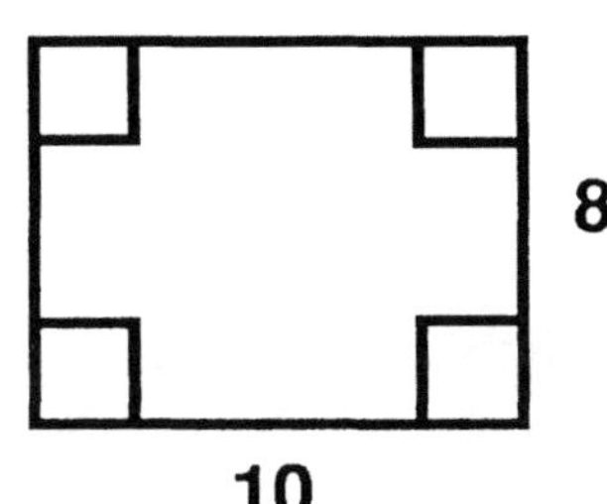

Two- by two-inch squares are cut from the corners of a 10- by 8-inch rectangular piece of metal, which is then folded to form an open box. What is the volume of this box in cubic inches?

(A) 18 (B) 48 (C) 80 (D) 96 (E) 160

❹ Which of the following is determined by division?

I. Price of a Mercedes if it is six times that of a Volkswagen
II. Difference in temperature between Dallas and St. Paul
III. Number of yards in 42 feet

(A) I (B) II (C) III (D) I and II (E) I and III

© Dale Seymour Publications

DO NOW

❶ If Jake earns $3.85 per hour, what is his salary for a 40-hour week?

(A) $114.00 (B) $104.00 (C) $114.40

(D) $104.40 (E) $154.00

❷ Select the smallest of the following tool sizes.

(A) 1/2 in. (B)15/32 in. (C) 33/64 in.

(D) 7/16 in. (E) 3/8 in.

❸ At the Shiffer dairy it takes 90 seconds to fill 15 one-gallon jugs of milk. How many minutes does it take to fill 90 jugs of milk?

(A) 2 (B) 6 (C) 7 (D) 8 (E) 9

❹

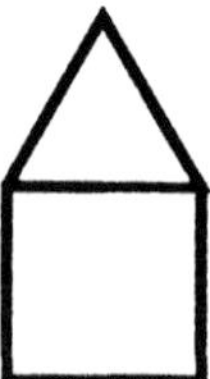

In the above figure, the equilateral triangle has a perimeter of 36. What is the area of the square?

(A) 36 (B) 48 (C) 72 (D) 144 (E) 324

© Dale Seymour Publications

DO NOW

QUANTITATIVE COMPARISONS:

Each question below consists of two quantities, one in column A and one in column B. (In some cases, additional information appears centered over the two columns.) You are to compare the two quantities and answer

- A if the quantity in Column A is greater;
- B if the quantity in Column B is greater;
- C if the two quantities are equal;
- D if the relationship cannot be determined from the information given.

	COLUMN A	COLUMN B	
❶	(0.82) (3.6)	(0.036) (8.2)	A B C D
❷	The average weight of Carlos, Mia, and Justin is 32 pounds.		
	The combined weight of Mia and Justin	Carlos's weight	A B C D
❸	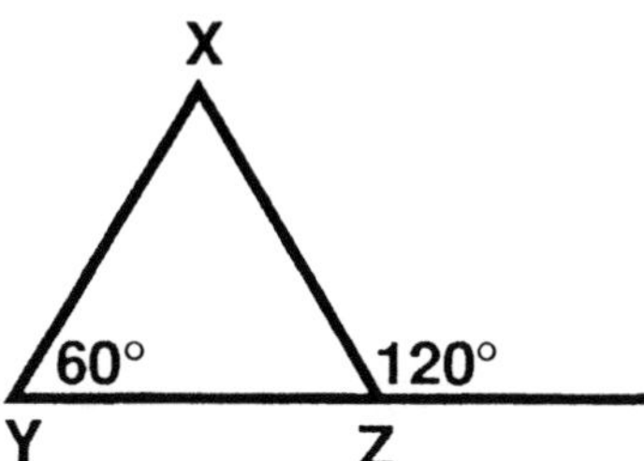		
	length of XY	length of YZ	A B C D
❹	$a^2 = 100$ $b^2 = 81$		
	a	b	A B C D
❺	4x	–4x	A B C D

© Dale Seymour Publications

DO NOW

1. Round off 253.0746 to the nearest hundredth.

(A) 253.07 (B) 253.08 (C) 253.1 (D) 2.53 (E) 253.075

2. If 9% of a number is 6.3, what is the number?

(A) 0.567 (B) 56.7 (C) 7 (D) 70 (E) 700

3. If Alex eats 1/4 of a pie, Mindy eats 1/3 of what's left after Alex is done, and Travis eats 1/2 of what's left after Mindy is done, then what part of the pie is left?

(A) 1/4 (B) 1/9 (C) 1/12 (D) 1/24 (E) none

4. Let x* be defined as the greatest integer that is less than or equal to x. (For example, 5.8* = 5 and 6* = 6.) Then –2.8* + 7* =

(A) 10 (B) 6 (C) 5 (D) 4 (E) 3

© Dale Seymour Publications

DO NOW

❶ Divide 2.68 by 100.

(A) 268 (B) 0.268 (C) 26.8 (D) 0.0268 (E) 0.00268

❷ 16 is 2/7 of what number?

(A) 4 4/7 (B) 14 (C) 28 (D) 56 (E) 112

❸ To the nearest thousand, what is the number of seconds in 12 hours?

(A) 4000 (B) 4500 (C) 43,000 (D) 44,000 (E) 45,000

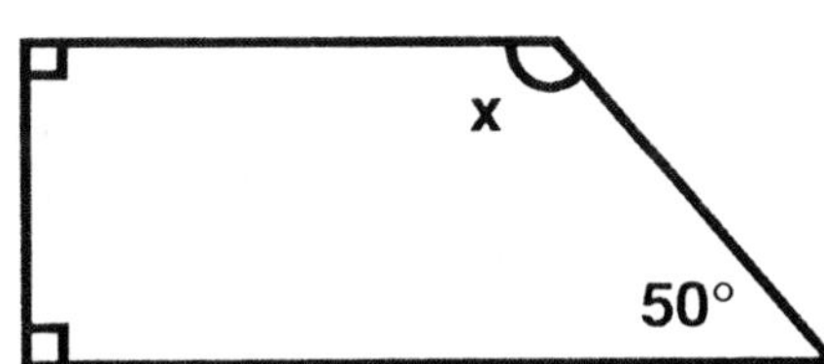

Find the measure of angle x.

(A) 40° (B) 90° (C) 130° (D) 180° (E) 360°

© Dale Seymour Publications

1. $6 + \frac{6}{0.6} =$

(A) 6.1 (B) 6.11 (C) 6.25 (D) 6 (E) 16

2. On the last day of a one-week sale, customers 359 through 401 were waited on. How many customers were waited on that day?

(A) 41 (B) 42 (C) 43 (D) 142 (E) 143

3. If the average of two numbers is 9 and the product of the two numbers is 80, then the positive difference between the two numbers is

(A) 1 (B) 2 (C) 3 (D) 4 (E) 6

4. Of the 60 people in a room, 2/3 are men, and 3/5 of the people have brown hair. What is the least number of men in the room who could have brown hair?

(A) 4 (B) 16 (C) 20 (D) 36 (E) 40

© Dale Seymour Publications

DO NOW

QUANTITATIVE COMPARISONS:

Each question below consists of two quantities, one in column A and one in column B. (In some cases, additional information appears centered over the two columns.) You are to compare the two quantities and answer

- A if the quantity in Column A is greater;
- B if the quantity in Column B is greater;
- C if the two quantities are equal;
- D if the relationship cannot be determined from the information given.

	COLUMN A	COLUMN B	
1	5^2	2^5	A B C D
2	0.02	(0.2) (10)	A B C D

3 The distance from Three Lakes to Eagle River is 9 miles. The distance from Eagle River to Rhinelander is 12 miles.

	COLUMN A	COLUMN B	
	21	The distance from Three Lakes to Rhinelander	A B C D

3 Two standard six-sided dice are rolled.

	COLUMN A	COLUMN B	
	Probability that the sum is 7	Probability that the sum is 8	A B C D

5

$a^2 = 9$

$b^2 = 16$

	COLUMN A	COLUMN B	
	a	b	A B C D

© Dale Seymour Publications

1. Write as a decimal fraction: $3 \frac{54}{10{,}000}$

(A) 3.54 (B) 3.054 (C) 3.0054 (D) 0.354 (E) 0.00354

2. In a class of 25 students, 10 are boys. If 5 more boys are admitted, what part of the class will be boys?

(A) 2/5 (B) 7/16 (C) 1/2 (D) 9/16 (E) 3/5

3. You have three different colored blocks—red, white, and blue. How many different ways could you stack them vertically?

(A) 3 (B) 4 (C) 6 (D) 8 (E) 9

4. A clock (not digital) loses 5 minutes a day. In how many days will it return to the correct time?

(A) 12 (B) 48 (C) 144 (D) 288 (E) 480

© Dale Seymour Publications

DO NOW

❶

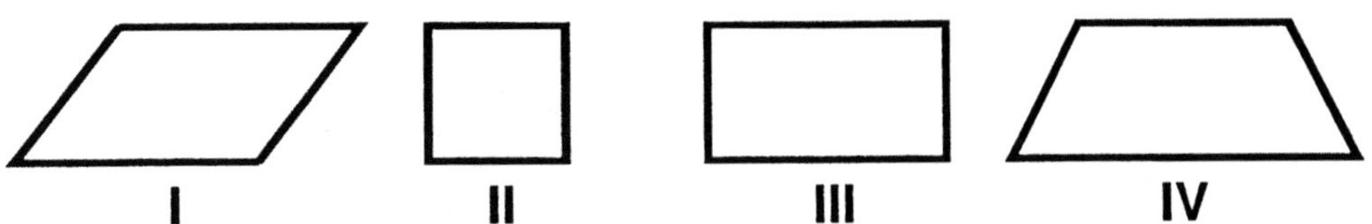

Which of the figures appear to be parallelograms?

(A) I (B) II and III (C) I, II, III (D) I and IV (E) none

❷ Five cards are picked from a standard deck of playing cards. What is the probability that all five cards are aces?

(A) 0 (B) 5/52 (C) 1/13 (D) 1/26 (E) 1

❸

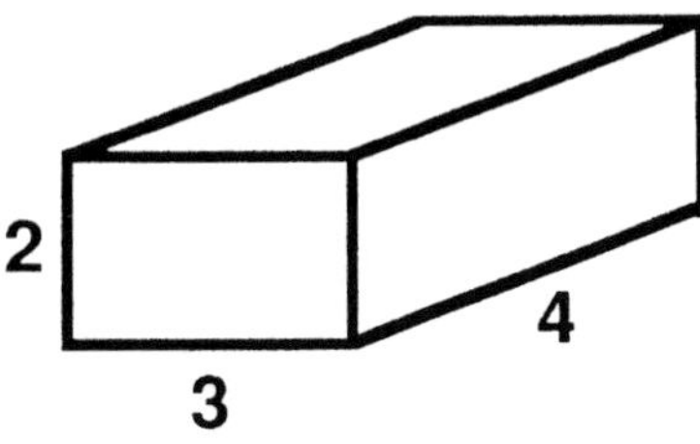

How many cubic units are there in the above figure?

(A) 9 (B) 14 (C) 24 (D) 30 (E) 144

❹ What part of 2 hours is 5 seconds?

(A) 5/7200 (B) 5/3600 (C) 5/120 (D) 5/2 (E) none of these

© Dale Seymour Publications

1. (3 1/2)(1/5) ÷ (1 2/5) =

(A) 2/5 (B) 1/2 (C) 49/50 (D) 50/49 (E) 2

2. Claudia went to the post office to mail 39 wedding invitations. She discovered that each invitation cost 42 cents to mail. How many could she mail if she had $8.50?

(A) 12 (B) 16 (C) 20 (D) 31 (E) 39

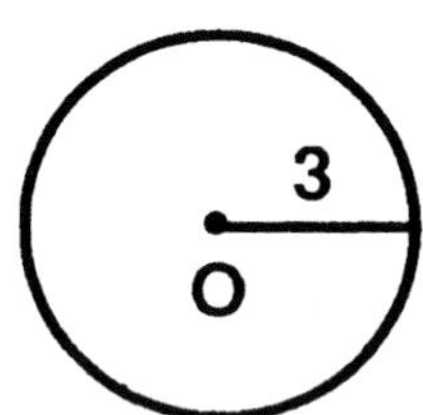

If point O is the center of the above circle, what is the approximate circumference of the circle?

(A) 6 (B) 10 (C) 19 (D) 28 (E) cannot determine

4. Six consecutive even integers are given. The sum of the first three is 24. What is the sum of the last three?

(A) 24 (B) 36 (C) 42 (D) 46 (E) none of these

© Dale Seymour Publications

DO NOW

QUANTITATIVE COMPARISONS:

Each question below consists of two quantities, one in column A and one in column B. (In some cases, additional information appears centered over the two columns.) You are to compare the two quantities and answer

- A if the quantity in Column A is greater;
- B if the quantity in Column B is greater;
- C if the two quantities are equal;
- D if the relationship cannot be determined from the information given.

	COLUMN A	COLUMN B	
❶	8	$\frac{2}{0.25}$	A B C D
❷	30% of x is 6.		
	x	18	A B C D
❸	(figure: quadrilateral with angles $a°$, $60°$, $60°$, $b°$)		
	$a°$	$b°$	A B C D
❹	$y < 0$ and $x > 0$		
	$y - x$	$x - y$	A B C D
❺	The number of sides on a hexagon	The number of sides on an octagon	A B C D

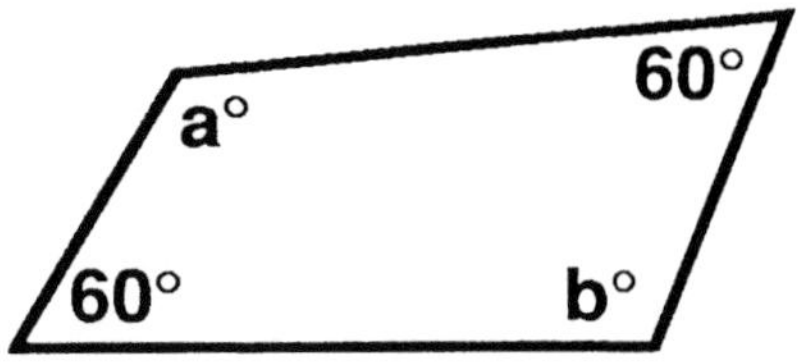

© Dale Seymour Publications

DO NOW

❶

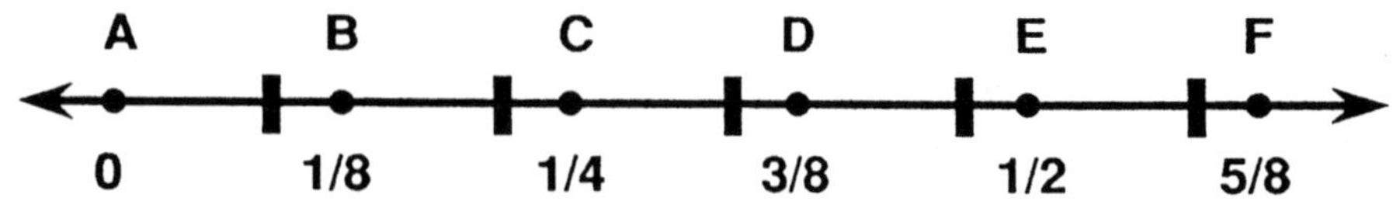

On the number line above, which letter best locates the number 0.476?

(A) A (B) B (C) C (D) D (E) E

---

❷ Combine 1/2 + 2/5 – 1/6.

(A) 1/15 (B) 4/15 (C) 11/15 (D) 14/15 (E) 29/15

---

❸ On his diet, Tim's weight dropped from 200 to 168 pounds. What was the percent decrease in his weight?

(A) 19% (B) 84% (C) 8.4% (D) 16% (E) 32%

---

❹

❖✿✿❑❑❑❖✿✿❑❑❑❖✿

Which of the following will best continue the above pattern?

(A) ❑❑ (B) ✿✿ (C) ❑✿ (D) ✿❑ (E) ❖❑

© Dale Seymour Publications

❶ Simplify: $7 \div 0.7/5$

(A) 0.5 (B) 5 (C) 50 (D) 2 (E) 0.2

❷ If a color television regularly sold at $495 is marked down 20%, what is the sale price?

(A) $119 (B) $396 (C) $405 (D) $483.20 (E) $475

❸ You flip a coin twice. (One side is heads; the other is tails.) What is the probability that you will get at least one head?

(A) 0 (B) 1/4 (C) 1/2 (D) 3/4 (E) 1

❹ At Crespi High School, 10 students are in the Science Club and 15 students are in the Math Club. There are 21 total students involved in the two clubs, as some students are members of both clubs. How many students are members of the Math Club only?

(A) 4 (B) 6 (C) 11 (D) 15 (E) 25

© Dale Seymour Publications

DO NOW

❶ Divide 56 by 0.035.

(A) 1.6 (B) 16 (C) 160 (D) 1600 (E) 16,000

❷ (5 + 6) ÷ 2/3 =

(A) 16 1/2 (B) 11 2/3 (C) 7 1/3 (D) 3/22 (E) 2/33

❸ If each of the 12 players on a basketball squad plays for the same amount of time, how many minutes does each play in a 48-minute game? (Five players play for a team at any one time.)

(A) 4 (B) 10 (C) 20 (D) 25 (E) 40

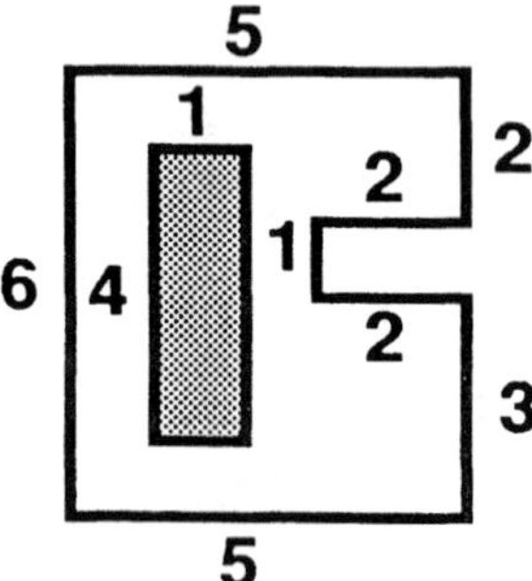

What is the area of the unshaded portion of the above figure?

(A) 34 (B) 30 (C) 28 (D) 26 (E) 24

© Dale Seymour Publications

DO NOW

QUANTITATIVE COMPARISONS:

Each question below consists of two quantities, one in column A and one in column B. (In some cases, additional information appears centered over the two columns.) You are to compare the two quantities and answer

- A if the quantity in Column A is greater;
- B if the quantity in Column B is greater;
- C if the two quantities are equal;
- D if the relationship cannot be determined from the information given.

	COLUMN A	COLUMN B	
1	$\frac{38.74}{2.7}$	$\frac{3874}{270}$	A B C D

2

$x > y > 0$

	COLUMN A	COLUMN B	
	1/x	1/y	A B C D

3

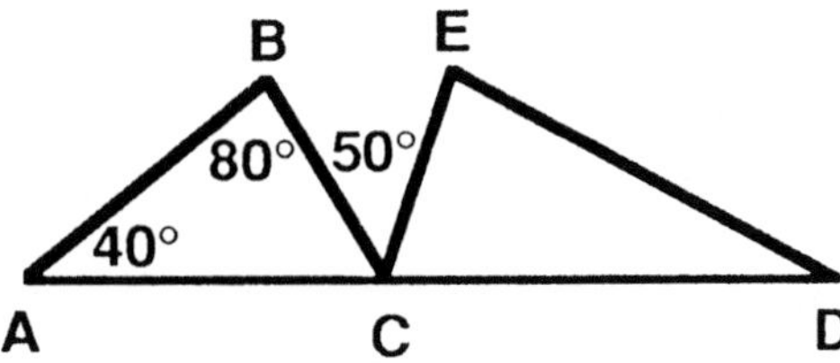

	COLUMN A	COLUMN B	
	Measure of angle ACB	Measure of angle ECD	A B C D

4

A pair of standard dice are rolled.

	COLUMN A	COLUMN B	
	The probability that a sum of 7 is rolled	The probability that a sum of 11 is rolled	A B C D

© Dale Seymour Publications

DO NOW

❶

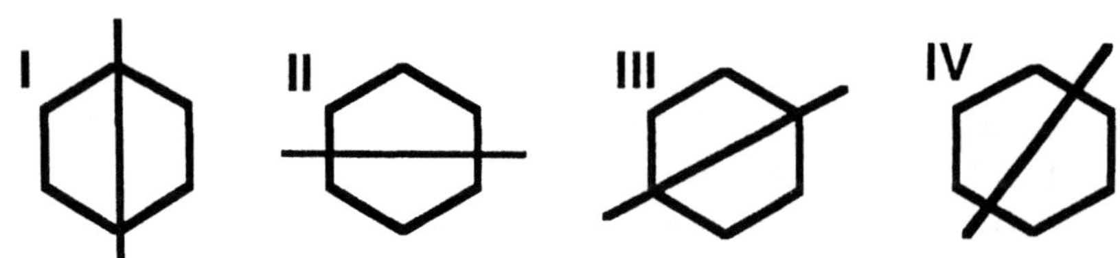

Which of the figures above shows a line of symmetry?
(All six sides of the figure are the same length.)

(A) I and III (B) II and IV (C) I, II, III

(D) all of them (E) none of them

❷ A car averages 34.8 miles per gallon and has a gas tank with a 12.2-gallon capacity. Of the following, which is the best approximation of the number of miles the car can be driven on a full tank of gasoline?

(A) 250 (B) 400 (C) 500 (D) 600 (E) 625

❸ Which of the following fractions has the greatest value?

(A) 25/52 (B) 31/60 (C) 19/40 (D) 51/103 (E) 43/90

❹ Find the missing term in the following set:

20, 25, 32, 41, 52, ____

(A) 55 (B) 61 (C) 63 (D) 65 (E) 80

© Dale Seymour Publications

DO NOW

 If cassette tapes cost $3.98 for a package of three, how much change will Kay get from a 20-dollar bill if she purchases twelve tapes?

(A) $2.08 (B) $3.08 (C) $4.08 (D) $5.08 (E) $15.92

❷ By how much does 1/2 of 5/6 exceed 1/3?

(A) 5/36 (B) 1/12 (C) 1/3 (D) 5/4 (E) 1/5

❸

Which of the following will best continue the above pattern?

(A) ❍▲ (B) ❍▼ (C) ✧▼ (D) ✧▲ (E) ❍❍

❹ Jay and Dixie were recently married and decided they wanted two children. What is the probability that they will have two boys?

(A) 1/4 (B) 1/3 (C) 1/2 (D) 1 (E) 0

© Dale Seymour Publications

DO NOW

1. Express 0.4% as a decimal.

(A) 0.4 (B) 4 (C) 0.004 (D) 0.0004 (E) 0.04

2. 124 is 20 percent of what number?

(A) 24.8 (B) 248 (C) 620 (D) 1240 (E) 2480

3. An insurance company calculates the value of a motorcycle by deducting each year 25% of the value of the motorcycle at the beginning of that year. What is the value of a two-year-old motorcycle that cost $4000 when new?

(A) $2000 (B) $2125 (C) $2250 (D) $2485 (E) $2500

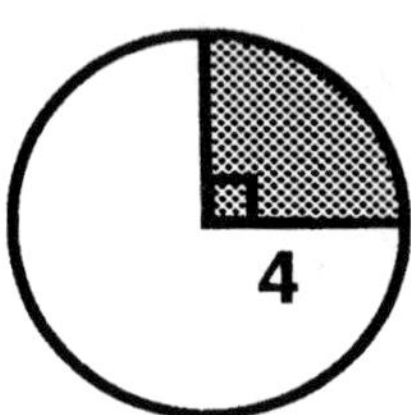

Find the area (in square units) of the shaded region in the above figure.

(A) 16π (B) 4π (C) 8π (D) 10π (E) 12π

© Dale Seymour Publications

DO NOW

QUANTITATIVE COMPARISONS:

Each question below consists of two quantities, one in column A and one in column B. (In some cases, additional information appears centered over the two columns.) You are to compare the two quantities and answer

- A if the quantity in Column A is greater;
- B if the quantity in Column B is greater;
- C if the two quantities are equal;
- D if the relationship cannot be determined from the information given.

	COLUMN A	COLUMN B	
❶	$\frac{0.9}{2}$	0.045	A B C D
❷	13% of 1500	15% of 1300	A B C D

❸ $x = 3$; $y = -2$

COLUMN A	COLUMN B	
$(x + y)^2$	$(x - y)^2$	A B C D

In the first five basketball games, Don scored 20, 13, 14, 9, and 19 points.

COLUMN A	COLUMN B	
Don's median score	Don's mean score	A B C D

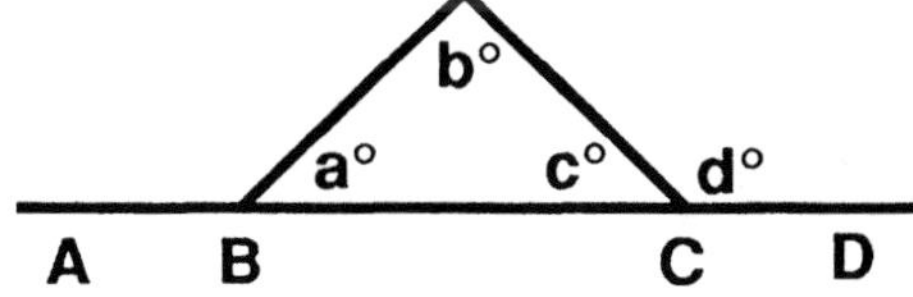

COLUMN A	COLUMN B	
$a° + b° + c°$	$c° + d°$	A B C D

© Dale Seymour Publications

DO NOW

1. What is 25% of 20.4?

(A) 5.1 (B) 51 (C) 510 (D) 81.6 (E) 816

2. A television marked $384 is offered for $288. The percent discount is

(A) 4% (B) 24% (C) 25%

(D) 33 1/3% (E) 66 2/3%

3. A bag contains 3 red marbles, 3 white marbles, and 4 blue marbles. If you pick one marble without looking, what is the probability that it is not blue?

(A) 3/10 (B) 2/5 (C) 3/5 (D) 7/10 (E) 1/2

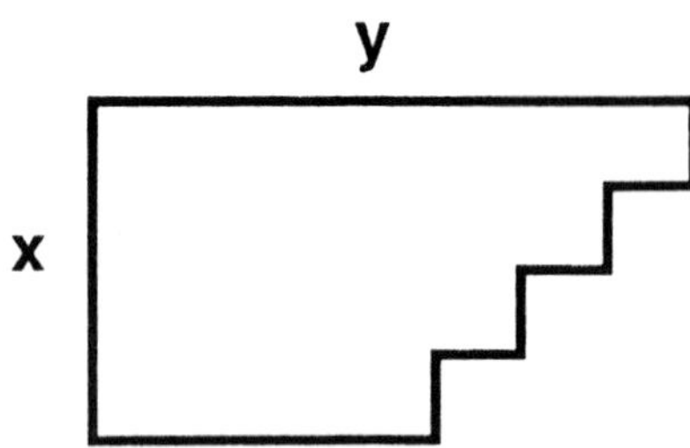

What is the perimeter of the above figure?

(A) $x + y$ (B) $x^2 + y^2$ (C) xy (D) $2(x + y)$ (E) none of these

© Dale Seymour Publications

DO NOW

1. If it costs 72 cents to make a calculator case, how many such cases can be produced for $65.00?

(A) 9 (B) 90 (C) 91 (D) 92 (E) 100

2. A board 7 feet 9 inches long is cut into three equal parts. The length of each part is

(A) 2' 5" (B) 2' 6" (C) 2' 7" (D) 2' 8" (E) 2' 9"

3. A pair of dice is rolled. What is the probability that the sum is 5?

(A) 1/36 (B) 1/18 (C) 1/9 (D) 1/6 (E) 1/4

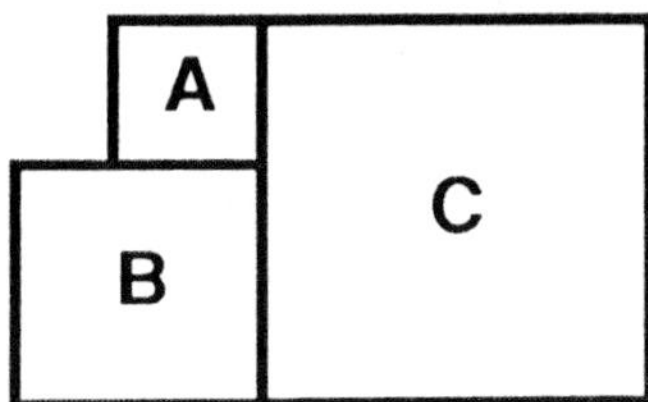

If the area of square A is 25 and the perimeter of square B is 32, find the perimeter of square C.

(A) 28 (B) 57 (C) 132 (D) 169 (E) none of these

© Dale Seymour Publications

DO NOW

❶ $(0.8)^3$ =

(A) 512 (B) 51.2 (C) 5.12 (D) 0.512 (E) 2.4

---

❷ Which of the following would give the closest approximation for the product of 35 x 45?

(A) 40 x 50 (B) 35 x 50 (C) 30 x 50 (D) 30 x 40 (E) 20 x 30

---

❸ A pair of dice are rolled. What is the probability that the sum will be 13?

(A) 1 (B) 1/36 (C) 1/18 (D) 13/36 (E) 0

---

❹ Let x* be defined as one more than the number of digits in the integer x. For example, 150* = 3 + 1 = 4. If x has 1000 digits, then what is the value of (x*)*?

(A) 1000 (B) 4 (C) 5 (D) 6 (E) 1001

© Dale Seymour Publications

DO NOW

QUANTITATIVE COMPARISONS:

Each question below consists of two quantities, one in column A and one in column B. (In some cases, additional information appears centered over the two columns.) You are to compare the two quantities and answer

- A if the quantity in Column A is greater;
- B if the quantity in Column B is greater;
- C if the two quantities are equal;
- D if the relationship cannot be determined from the information given.

	COLUMN A		COLUMN B	
❶	$\frac{0.3}{60}$		0.05	A B C D
❷		$a + b = 5$		
	a		b	A B C D
❸		$x = 6$ and $y = 1/3$		
	$2x - 18y$		$3x - 36y$	A B C D
❹		X workers working at r rate complete a job in J days. Y workers working at r rate complete a job in J + 3 days.		
	X		Y	A B C D

❺

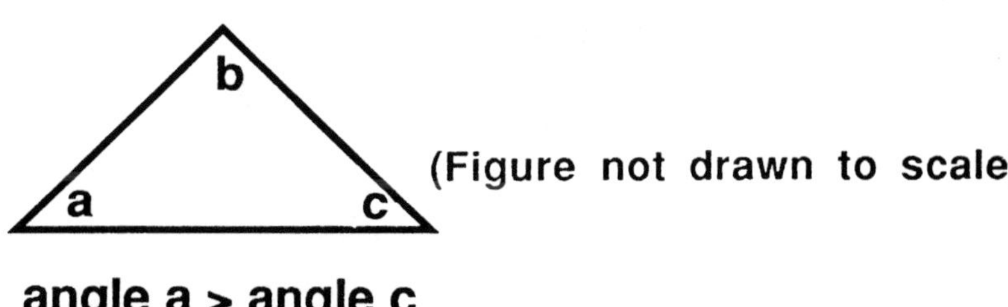

angle a > angle c
angle b < angle a

COLUMN A	COLUMN B	
angle b	angle c	A B C D

© Dale Seymour Publications

DO NOW

❶ $100 - \frac{100}{0.1} =$

(A) 900 (B) 990 (C) –900 (D) 0 (E) 90

---

❷ Points X and Y on line WZ are placed so that WX = XY = YZ. What percent of XZ is WZ?

(A) 50% (B) 66 2/3% (C) 150% (D) 200% (E) 300%

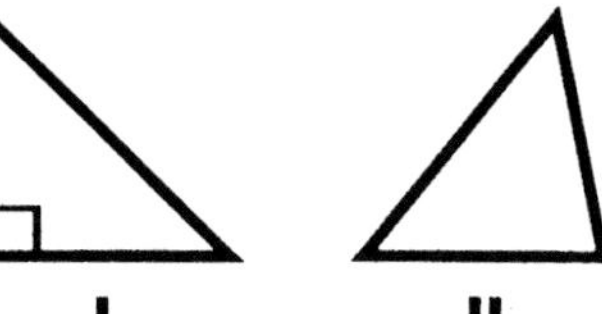

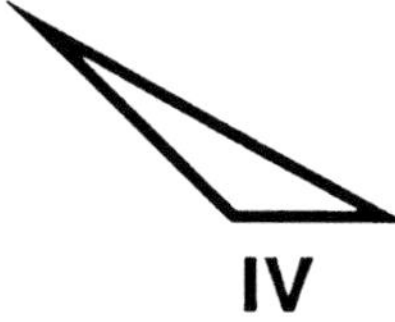

I II III IV

Which of the above triangles appear to be similar?

(A) I, II, IV (B) I and III (C) I and II

(D) all of them (E) none of them

---

❹ Initially there are exactly 36 oranges on a tree. If Luis eats 1/3 of the oranges and Madelyn eats 1/3 of the oranges that are left, how many oranges are still on the tree?

(A) 8 (B) 10 (C) 12 (D) 14 (E) 16

© Dale Seymour Publications

DO NOW

 Which of the following has the greatest value?

(A) 0.4044 (B) 0.44 (C) 0.404 (D) 0.04444 (E) 0.444

❷ By how much does 3/8 of 1 2/3 exceed 1/6 of 3?

(A) 5/16 (B) 1/8 (C) 5/4 (D) 4/5 (E) 16/5

❸

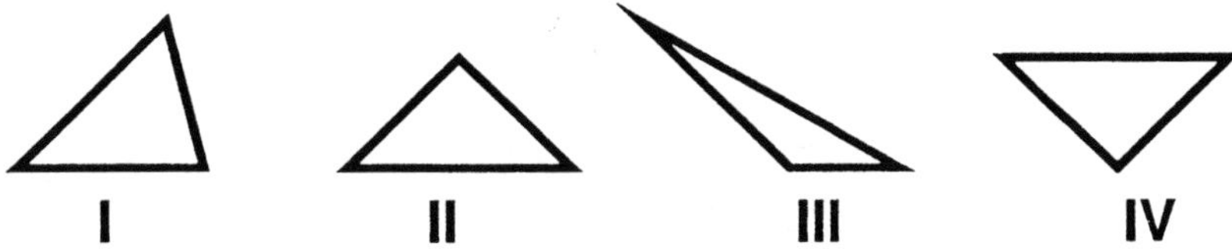

Which two triangles appear to be congruent?

(A) I and III (B) II and IV (C) I, II, III

(D) all of them (E) none of them

❹ A box contains different colored balls of the same size and shape: one green ball, two red balls, and three blue balls. Maria draws one ball from the box without looking. What is the probability that the ball will be either green or blue?

(A) 1/6 (B) 5/6 (C) 2/3 (D) 1/2 (E) 1/3

© Dale Seymour Publications

DO NOW

❶ What is the product of –9.6 and –2.7?

(A) –2.592 (B) 2.592 (C) –25.92 (D) 25.92 (E) 259.2

❷ A standard die is rolled 600 times. About how many times should it land on 4?

(A) 50 (B) 100 (C) 300 (D) 400 (E) 600

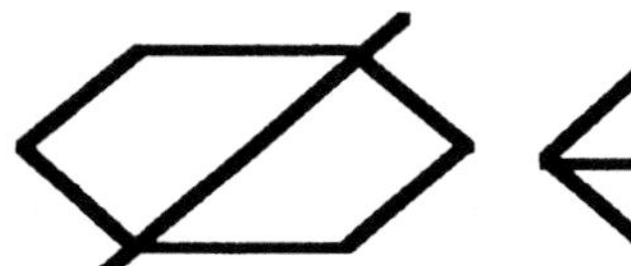   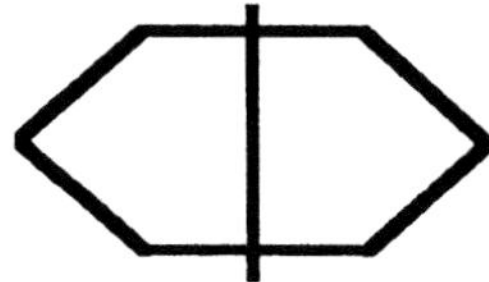

Which of the above figures show a line of symmetry?

(A) I and III (B) II and IV (C) I, II, III

(D) all of them (E) none of them

❹ If a wheel rotates 8 times each minute, how many degrees does it rotate in 15 seconds?

(A) 90° (B) 180° (C) 360° (D) 720° (E) 1080°

© Dale Seymour Publications

DO NOW

QUANTITATIVE COMPARISONS:

Each question below consists of two quantities, one in column A and one in column B. (In some cases, additional information appears centered over the two columns.) You are to compare the two quantities and answer

A if the quantity in Column A is greater;
B if the quantity in Column B is greater;
C if the two quantities are equal;
D if the relationship cannot be determined from the information given.

	COLUMN A	COLUMN B	
❶	$\frac{13}{21}$	$\frac{8}{21} + \frac{5}{23}$	A B C D ▯ ▯ ▯ ▯
❷	$(-5)^{16}$	$(-5)^{17}$	A B C D ▯ ▯ ▯ ▯
❸	The area of a square with a side of 5	The area of a triangle with a base of 5	A B C D ▯ ▯ ▯ ▯

❹ The price of a TV is increased by 25%.
The new price is decreased by 25%.

	COLUMN A	COLUMN B	
❹	The original price of the TV	The latest price of the TV	A B C D ▯ ▯ ▯ ▯

❺

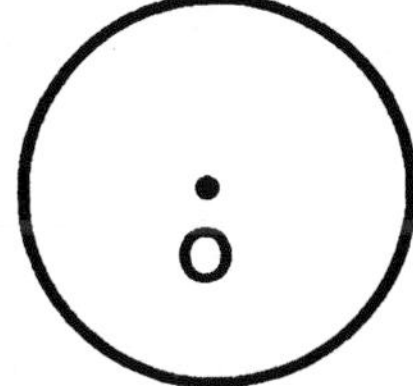

	COLUMN A	COLUMN B	
❺	Radius of circle O	Chord of circle O	A B C D ▯ ▯ ▯ ▯

© Dale Seymour Publications

DO NOW

1 The price of almonds dropped from \$5.25 a pound to \$4.20 a pound. What was the percent decrease?

(A) 20% (B) 23% (C) 25% (D) 27% (E) 30%

2 For all positive integers n, let

$$n^* = n + 1 \text{ if } n \text{ is even;}$$
$$n^* = n - 1 \text{ if } n \text{ is odd.}$$

Then $3^* \cdot 8^* =$

(A) 14 (B) 18 (C) 24 (D) 28 (E) 36

3 If $\frac{5}{6} \cdot \frac{6}{7} \cdot \frac{7}{8} \cdot \frac{8}{9} \cdot \frac{9}{10} \cdot x = 1$,

then what is the value of x?

(A) 10/11 (B) 1 (C) 2 (D) 10/9 (E) 1/9

4 Molly kept a record of her base hits. She has 20 hits in her last 48 at bats. Based on her record, what is the probability that she will get a hit her next time at bat?

(A) 2/5 (B) 1/3 (C) 5/12 (D) 1/2 (E) 1

© Dale Seymour Publications

DO NOW

1 100 is what percent of 40?

(A) 20% (B) 40% (C) 100% (D) 140% (E) 250%

2 A standard die is rolled 600 times. About how many times should a 4 or a 5 appear?

(A) 50 (B) 100 (C) 200 (D) 300 (E) 400

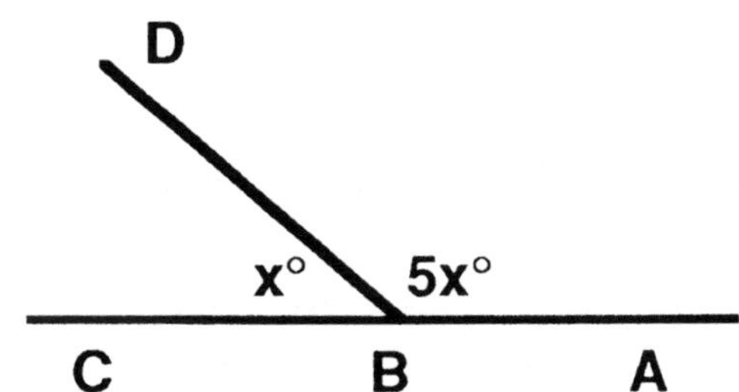

Assuming points A, B, and C are colinear, what is the measure in degrees of angle ABD?

(A) 30° (B) 60° (C) 90° (D) 120° (E) 150°

4 In a group of 60 people, 42 have brown hair, 38 have brown eyes, and 28 have both brown hair and brown eyes. How many have neither brown hair nor brown eyes?

(A) 8 (B) 10 (C) 14 (D) 24 (E) 52

© Dale Seymour Publications

DO NOW

❶ How many fourths are there in 66 2/3%?

(A) 1/6 (B) 3/8 (C) 2 2/3 (D) 3 (E) 4

❷ Jason will be x years old 6 years from now. How old was he 6 years ago?

(A) x – 6 (B) x + 12 (C) x – 12 (D) 6x – 6 (E) 6x

❸ A number from 1 to 100 is picked at random. What is the probability that it is a multiple of 6?

(A) 1/10 (B) 3/25 (C) 7/50 (D) 4/25 (E) 9/50

❹ Let x* be defined as one more than the number of digits in the integer x. For example, 138* = 3 + 1 = 4. If x is a positive integer less than one million, then x* is, at most,

(A) 6 (B) 7 (C) 8 (D) 9 (E) 1 million

© Dale Seymour Publications

DO NOW

QUANTITATIVE COMPARISONS:

Each question below consists of two quantities, one in column A and one in column B. (In some cases, additional information appears centered over the two columns.) You are to compare the two quantities and answer

- A if the quantity in Column A is greater;
- B if the quantity in Column B is greater;
- C if the two quantities are equal;
- D if the relationship cannot be determined from the information given.

	COLUMN A	COLUMN B	
1	55☐ is a three-digit number divisible by 9.		
	☐	7	A B C D ▯ ▯ ▯ ▯
2	$(0.456)^2$	0.456	A B C D ▯ ▯ ▯ ▯
3	A circle has a radius of 2.		
	Area of the circle	Circumference of the circle	A B C D ▯ ▯ ▯ ▯

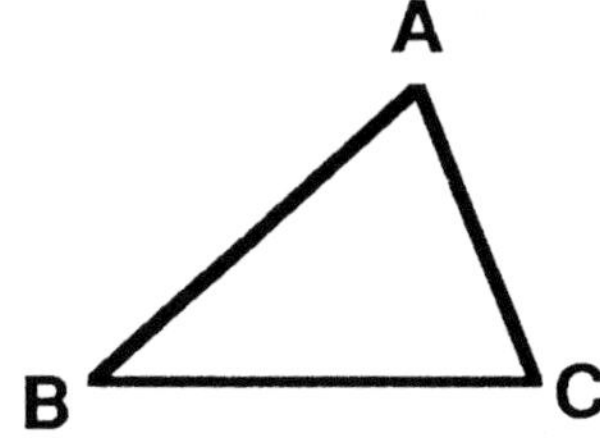

Triangle ABC is isosceles. AB = BC.

COLUMN A	COLUMN B	
$\frac{AC}{AB}$	$\frac{AC}{BC}$	A B C D ▯ ▯ ▯ ▯

x * y means $(x - y)^2$

COLUMN A	COLUMN B	
4 * 6	2 * –1	A B C D ▯ ▯ ▯ ▯

© Dale Seymour Publications

DO NOW

❶ A man buys a $240 stereo at a discount of 25%. How much does he pay for it?

(A) $160 (B) $180 (C) $192 (D) $200 (E) $208

❷ An eight-sided die (numbered 1 through 8) is rolled 200 times. About how many times should it land on 5?

(A) 8 (B) 13 (C) 25 (D) 40 (E) 100

❸ Heather keeps a record of her ace serves in tennis. If she has served 12 aces in her previous 54 attempts, what is the probability that she will serve an ace on her next attempt?

(A) 1/4 (B) 2/9 (C) 1/3 (D) 1/2 (E) 2/7

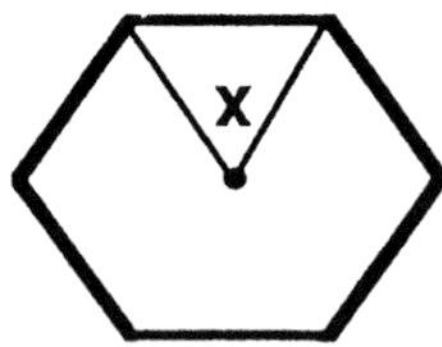

Find the measure of angle x for the regular (equal-sided) hexagon.

(A) 30° (B) 60° (C) 90° (D) 120° (E) cannot determine

© Dale Seymour Publications

DO NOW

❶ If the temperature in Green Bay was −17.5° yesterday and +5.6° today, how much did the temperature rise?

(A) 1.19° (B) 2.31° (C) 11.9° (D) 23.1° (E) 33.1°

❷ You flip a coin three times. (One side is heads; the other is tails.) What is the probability that you will get all three flips to land tails?

(A) 0 (B) 1/8 (C) 1/4 (D) 1/2 (E) 3/4

❸ The Osawas purchased 18 square yards of carpeting for their house. How many square feet of carpet is this?

(A) 2 (B) 6 (C) 54 (D) 162 (E) cannot determine

❹ Let x* be defined as the greatest integer that is less than or equal to x. For example, 5.8* = 5; 6* = 6. Then 1.4* + 3.9* =

(A) 4 (B) 5 (C) 6 (D) 7 (E) 8

© Dale Seymour Publications

DO NOW

❶ How many gallons of water are there in 12 gallons of a solution that is 10% pure acid, with the remainder being water?

(A) 1.2 (B) 2.4 (C) 4.8 (D) 9.6 (E) 10.8

❷ If line AB touches a circle only at point C, what term describes the relationship between line AB and the circle?

(A) tangent (B) complementary (C) chord (D) parallel (E) perpendicular

❸ If there are exactly two different types of cars and exactly three different colors of paint, then how many different car/color combinations are there?

(A) 2 (B) 3 (C) 5 (D) 6 (E) 9

❹ Ramon was k years of age 2 years ago. In terms of k, how old will he be 2 years from now?

(A) k + 4 (B) k + 2 (C) 2k (D) k (E) k/2

© Dale Seymour Publications